The Future of Computing: Ubiquitous Applications and Technologies

Edited by

Neha Kishore

Department of Computer science and Engineering
Maharaja Agrasen Institute of Technology
Maharaja Agrasen University
Himachal Pradesh, India

Pankaj Nanglia

Department of Electronics and Engineering
Maharaja Agrasen Institute of Technology
Maharaja Agrasen University
Himachal Pradesh, India

Shilpa Gupta

Department of Electronics and Communication Engineering
Chandigarh University
Mohali, India

&

Ashutosh Kumar Dubey

Department of Computer Science and Engineering
Chitkara University
Himachal Pradesh, India

The Future of Computing: Ubiquitous Applications and Technologies

Editors: Neha Kishore, Pankaj Nanglia, Shilpa Gupta & Ashutosh Kumar Dubey

ISBN (Online): 978-981-5238-99-0

ISBN (Print): 978-981-5256-00-0

ISBN (Paperback): 978-981-5256-01-7

First published in 2024.

need for a court order if at any point you breach any terms of this License Agreement. In no event will any delay or failure by Bentham Science Publishers in enforcing your compliance with this License Agreement constitute a waiver of any of its rights.

3. You acknowledge that you have read this License Agreement, and agree to be bound by its terms and conditions. To the extent that any other terms and conditions presented on any website of Bentham Science Publishers conflict with, or are inconsistent with, the terms and conditions set out in this License Agreement, you acknowledge that the terms and conditions set out in this License Agreement shall prevail.

Bentham Science Publishers Pte. Ltd.
80 Robinson Road #02-00
Singapore 068898
Singapore
Email: subscriptions@benthamscience.net

CONTENTS

PREFACE

The Future of Computing: Ubiquitous Applications and Technologies delves into the exciting world of ubiquitous computing and its diverse applications across various domains. Ubiquitous computing refers to the concept of seamlessly integrating computing technologies into our everyday lives, making them pervasive and invisible.

In this book, we explore the potential of ubiquitous computing in addressing critical challenges and revolutionizing different sectors. The chapters presented here offer a comprehensive overview of cutting-edge research and practical implementations, providing valuable insights into researchers, practitioners, and enthusiasts alike.

"Automated Analysis of Medical Images for Healthcare Domain," sheds light on the advancements in medical imaging analysis, leveraging the power of ubiquitous computing. The chapter explores how automated techniques can improve healthcare outcomes, facilitate diagnoses, and enhance patient care. The chapter titled, "Towards the Assessment of Federations of Clouds," examines the emerging trend of federated clouds and their implications. It discusses the challenges, benefits, and potential applications of federated cloud environments, showcasing how ubiquitous computing can optimize cloud-based services. The chapter, "Digital Payments & Financial Cyber Frauds in Rural India," explores the intersection of ubiquitous computing and financial inclusion in rural India. This chapter investigates digital payment systems and the challenges posed by financial cyber frauds, presenting potential solutions to enhance security and promote safe digital transactions. Another chapter, "Speed Control of DC Motor using PID Controller with Artificial Intelligence Techniques," delves into the realm of control systems and artificial intelligence. It showcases how ubiquitous computing, coupled with PID controllers and AI techniques, can optimize the performance of DC motors, enabling precise speed control. "Power System Harmonic Analysis and Elimination," focuses on the application of ubiquitous computing in power systems. It delves into the challenges posed by harmonics and explores advanced techniques for harmonic analysis and elimination, ultimately enhancing the reliability and efficiency of power grids. The next chapter, "AutoMate: Ubiquitous Smart Home System using Arduino and ESP8266 Module," presents an innovative approach to home automation. By leveraging ubiquitous computing technologies such as Arduino and ESP8266, the chapter demonstrates the development of a smart home system that seamlessly integrates devices and enhances user convenience. The next chapter, "Digital Forensics in Mobile Phones: An Overview of Data Acquisition Techniques and its Challenges," delves into the realm of digital forensics in the context of ubiquitous mobile devices. It provides an overview of data acquisition techniques, challenges, and emerging trends in mobile forensics, highlighting the importance of ubiquitous computing in investigations. The chapter titled, "IoT and AIoT: Applications, Challenges, and Optimization," explores the convergence of the Internet of Things (IoT) and Artificial Intelligence of Things (AIoT). It investigates the applications, challenges, and optimization strategies in this rapidly evolving field, showcasing the transformative potential of ubiquitous computing. The chapter "IoT Semantic of AI Security Structure for Smart Grid," focuses on the application of ubiquitous computing in securing smart grids. It presents an in-depth analysis of the semantic aspects of IoT and AI security structures, highlighting the importance of robust security measures for critical infrastructure.

Throughout this book, we strive to provide an insightful exploration of ubiquitous computing's applications and challenges across various domains. By bringing together expert perspectives and cutting-edge research, we aim to inspire further innovation and advancement in this fascinating field. We hope that this book serves as a valuable resource, fostering a deeper understanding of ubiquitous computing and its limitless possibilities.

It gives us immense pleasure to express our gratitude to the individuals who have made significant contributions and provided valuable assistance throughout the creation of this book. We extend our heartfelt thanks to all the authors who submitted their chapters, as their contributions and insightful discussions have played a pivotal role in making this book a resounding success. We sincerely hope that readers will find great value and gain future insights from the diverse contributions made by these authors. Furthermore, this book serves as a catalyst, opening new avenues and opportunities for future research in the field of ubiquitous computing. We are deeply grateful to the dedicated team at Bentham Publication for their meticulous service and timely publication of this book, ensuring its availability to the wider audience. We would also like to extend our profound appreciation to our institutions/universities and colleagues for their unwavering support and encouragement throughout this endeavor. Their support has been instrumental in bringing this book to fruition.

Lastly, we would like to acknowledge and express our heartfelt gratitude to our families for their unwavering support, encouragement, and patience. Their understanding and belief in us have been a constant source of motivation. Once again, we extend our sincere thanks to all those who have contributed to the realization of this book. It is their collective efforts and support that have made this publication possible.

Neha Kishore
Department of Computer science and Engineering
Maharaja Agrasen Institute of Technology
Maharaja Agrasen University
Himachal Pradesh, India

Pankaj Nanglia
Department of Electronics and Engineering
Maharaja Agrasen Institute of Technology
Maharaja Agrasen University
Himachal Pradesh, India

Shilpa Gupta
Department of Electronics and Communication Engineering
Chandigarh University
Mohali, India

&

Ashutosh Kumar Dubey
Department of Computer Science and Engineering
Chitkara University
Himachal Pradesh, India

List of Contributors

Amit Verma	Maharaja Agrasen Institute of Technology, Maharaja Agrasen University, Himachal Pradesh, India
Bharat Chhabra	Department of Computer Science, Govt. College for Women, Karnal, India
Bindu Thakral	Sushant University, Gurugram, Haryana, India
Kandipati Rajani	Department of Electrical and Electronics Engineering, Vignan's Lara Institute of Technology and Sciences, Guntur, Andhra Pradesh, India
Neha Gupta	Department of Computer Science & Engineering, Institute of Engineering & Technology, Chitkara University, Rajpura, Punjab, India
Neeli Manoj Venkata Sai	Department of Electrical and Electronics Engineering, R.V.R. & J.C. College of Engineering, Guntur, Andhra Pradesh, India
Neha Kishore	Department of Computer Science and Engineering, Maharaja Agrasen University, Himachal Pradesh, India
Parul Chhabra	Department of Computer Science & Engineering, G. J. University of Science & Technology, Hisar, Haryana, India
Pradeep Kumar Bhatia	Department of Computer Science & Engineering, G. J. University of Science & Technology, Hisar, Haryana, India
Priya Raina	School of Engineering and Technology, Chitkara University, Himachal Pradesh, India
Ranjit Kumar	Department of Computer Science & Engineering, Maharaja Agrasen University, Baddi, Himachal Pradesh, India
Rahul Gupta	Department of Computer Science & Engineering, Maharaja Agrasen University, Baddi, Himachal Pradesh, India
Rahul Rajput	Sushant University, Gurugram, Haryana, India
Rama Koteswara Rao Alla	Department of Electrical and Electronics Engineering, R.V.R. & J.C. College of Engineering, Guntur, Andhra Pradesh, India
Rakhi Kamra	Department of Electrical and Electronics Engineering, Maharaja Surajmal Institute of Technology, Delhi, India
Raman Kumar	KGPTU, Kapurthala, Jalandhar, Punjab, India
Sunil Kumar	Guru Jambheshwar University of Science & Technology, Hisar, Haryana, India
Shilpa Gupta	Department of Electronics and Communication Engineering, Chandigarh University, Mohali, India
Soumya Chaudhary	Department of Electrical and Electronics Engineering, Maharaja Surajmal Institute of Technology, Delhi, India
Vipin Babbar	Department of Computer Science & Engineering, G. J. University of Science & Technology, Hisar, Haryana, India

CHAPTER 1

Automated Analysis of Medical Images in the Healthcare Domain

Parul Chhabra[1,*], Pradeep Kumar Bhatia[1] and Vipin Babbar[1]

[1] *Department of Computer Science & Engineering, G. J. University of Science & Technology, Hisar, Haryana, India*

Abstract: During lab tests, thousands of medical images are generated to trace the disease's symptoms. Manual interpretation of this data may consume excessive time and thus may delay diagnosis. Timely detection of critical diseases is very important as their stage can be changed over an interval. Automated analysis of medical data can reduce the gap between disease detection and its diagnosis and it also reduces the overall computational cost. In this paper, this goal will be achieved using different methods (Classification/ Segmentation/ Image Encoding/ Decoding/ Registration/ Restoration/ Morphology).

Keywords: Disease, Diagnosis, Healthcare, Medical image analysis, Prediction.

INTRODUCTION

Traditional healthcare services follow different steps *i.e.* disease detection, diagnosis, and keeping track of a patient's history for clinical decision-making, as shown in Fig. (**1**). Medical data produced by each step must be examined by expert practitioners to avoid the incorrect diagnosis.

The disease detection phase may produce a large set of medical images and precise analysis of these medical images plays an important role in the identification of disease. It can also be used to track the progress of diagnosis as well as different stages of disease w.r.t. patients.

---

[*] **Corresponding author Parul Chhabra:** Department of Computer Science & Engineering, G. J. University of Science & Technology, Hisar, Haryana, India; E-mail: parul15march@gmail.com

Neha Kishore, Pankaj Nanglia, Shilpa Gupta & Ashutosh Kumar Dubey (Eds.)

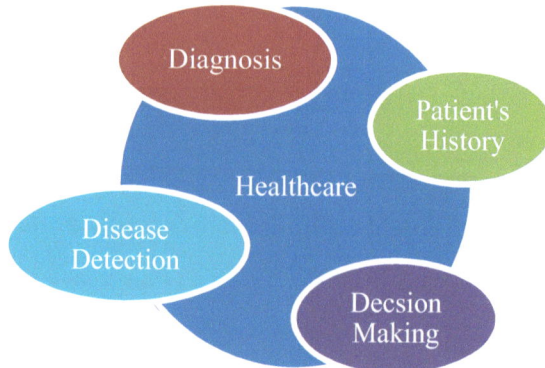

Fig. (1). Health care services.

Machine learning can improve the efficiency of the analysis process and it can also be used to build a dataset/knowledgebase for healthcare services in such a way that patient/disease statistics can be shared worldwide. Medical images contain data in visual form and only expert practitioners can interpret that data [1 - 25].

To analyze this data automatically, machine learning offers the following ways as displayed in Fig. (**2**).

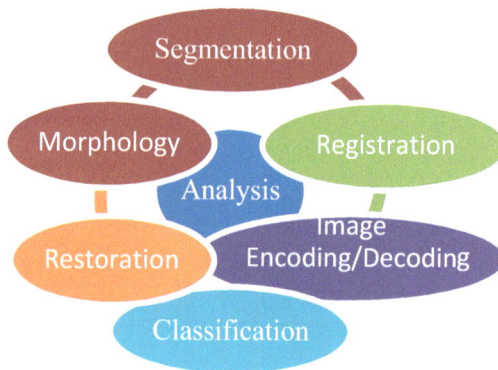

Fig. (2). Medical image analysis.

- Classification: Medical images can be classified w.r.t. disease types/features *etc.* and they can be used to detect disease and diagnostic purposes [26].
- Segmentation: It is used to subdivide an image into multiple segments (*i.e.* objects/regions). It can be used for pathologies domain/object detection/ recognition, *etc.* [27].

- Image Encoding/Decoding: It is used to compress the image whereas decoding follows the reverse operation to obtain the original image [28].
- Registration: It can be used to align and stitch multiple images together for analysis purposes [29].
- Restoration: It is used to filter noise level in an image, in order to produce clear and refined output [30].
- Morphology: It deals with structural components, pixels, and shapes in a given image [31].

Following are the challenges and limitations of automated medical image analysis:

- It requires a large volume of medical datasets, in order to build a training model for prediction.
- Dataset validation is required to ensure the accuracy of the training model.
- Quite complex to update the existing dataset.
- Excessive computational resources are required to manage and process large-scale medical data.
- Expert medical practitioners are still required to ensure the validity of outcomes.

The potential impact of automation of medical image analysis is given below:

- It can reduce the processing time and computational cost for practitioners.
- It can increase the accuracy of clinical decision-making.
- It can optimize the errors in the diagnosis process.
- Training model can be updated using the patient's history, and health recovery with respect to recommended treatment.

Researchers have developed a few solutions for the analysis of medical imagery as discussed in the next section.

LITERATURE SURVEY

K. Rasheed *et al*. [6] investigated the various machine learning (ML) applications for the healthcare domain. Studies found that intelligent solutions can improve the diagnosis accuracy however, there are a few open issues *i.e.* lack of standards to generate the training models, dataset formats, incompatible interfaces for the data exchange, *etc.*

R. Buettner *et al*. [7] highlighted the various ML-based methods that can be utilized for medical image processing *i.e.* medical image encoding/decoding, segmentation, classification, image registration/restoration, morphological analysis, *etc.* Study shows that the accuracy of disease detection can be improved using these methods.

D. Tellez *et al.* [8] developed an image compression method that encodes the histopathology dataset and uses neural networks to compress noise level input. Outcomes show that it can produce refined images with optimal reconstruction error and these images can be easily interpreted by practitioners.

P. Seeböck [9] introduced an ML-based method that uses supervised learning for the analysis of retina images. It enforces binary classification, noise filtering, and clustering over input data to detect anomalies. Experimental results show that it has an average accuracy/ROC curve.

K. Gong *et al.* [10] introduced an image reconstruction method that builds a learning model using neural networks. It estimates the energy levels (low/high) in a given input and uses multipliers to reconstruct the images. Experiments show that it is more efficient as compared to traditional denoising methods.

Y. Qi *et al.* [11] developed a neural network-based method to improve the quality of images. It estimates different parameters (contrast/coherence/signal-to-noise-ratio) to reproduce the high-resolution images. Analysis indicates that it is more efficient as compared to existing solutions.

Q. Abbas *et al.* [12] developed an ML model to analyze medical images. It builds metrics using various processes (segmentation/regression/regeneration/augmentation/loss function/data loading). Experiments indicate that these metrics can be used to enhance the diagnosis accuracy as well as reduce operational costs.

X. Zhou *et al.* [13] developed a neural network to classify histopathological data. It uses segmentation to produce outcomes and tests have shown that it is more accurate and efficient as compared to traditional deep-learning neural networks.

H. Guan *et al.* [14] conducted a survey to analyze the impact of different factors associated with medical images *i.e.* data type/volume/quality *etc.* Analysis shows that the accuracy of disease detection and decision-making is affected by learning methods (supervised/unsupervised/semi-supervised) and computational costs may vary due to heterogeneous data types.

X. Wang *et al.* [15] explored the relationship between image analysis and diagnostic accuracy and developed a classification model using supervised learning. It builds labeled data for each input to perform classification. Outcomes show that it has an optimal computational cost, higher accuracy, and efficiency in contrast with traditional methods.

H. Pinckaers *et al*. [16] used neural networks to improve the quality of image data. It extracts the Metadata of images and forms a correlation metric to produce high-resolution data. The analysis has shown that it can manage variations in the input size, pixel size, *etc.* and it has an optimal ROC value.

B. M. Rashed *et al*. [17] investigated different ML algorithms that can be used for data mining of medical images and these are Support Vector Machine (SVM)/ k-Nearest Neighbors (KNN)/Decision Trees/Random Forest/Logistic Regression. The common processes found in the study are prediction, data mining, classification, regression, clustering, dimension reduction, *etc.* that can be used for image analysis.

K. Naveen *et al*. [18] studied the role of machine learning/deep learning algorithms for the medical image analysis. It has been found that learning methods can affect the accuracy of analysis and the computational cost may also vary. This study states that the optimal selection of a learning method is necessary to ensure good outcomes in the classification process.

S. P. Shayesteh *et al*. [19] developed a method to extract features from ultrasound samples. It uses logistic regression classifier to perform selective feature selection. The analysis has shown that it offers optimal sensitivity with higher accuracy of disease detection in contrast to existing solutions.

T. Zhang *et al*. [20] identified that noise in images can degrade the outcomes of a classifier and introduced a noise adaption solution that enforces noise patterns over ultrasound samples. Outcomes show that it has higher accuracy under the constraints of noise level variations.

Geetha *et al*. [21] compared the performance of supervised/unsupervised learning approaches with respect to healthcare data. The analysis has shown that the accuracy of a prediction model may vary with respect to learning methods. Also, the complexity of medical data types may reduce its efficiency, which may also affect the decision making process and increase the operational cost.

N. Nahar *et al*. [22] explored the association between disease detection and the analysis of X-ray images. The study found that deep learning algorithms can efficiently predict the presence of diseases in given input samples and improve the diagnostic accuracy. However, the study also indicates that there is no single solution to analyze different types of medical imagery.

U. Khan *et al*. [23] found that ML algorithms can be used to extract the clinical data from medical imagery efficiently by using classification and segmentation techniques. The outcomes of the analysis can be further used to improve the accuracy of diagnosis and clinical decision-making.

A. Sivasangari *et al.* [24] developed a solution to identify brain tumor using neural networks. First of all, it subdivided the brain cells into two different categories (healthy/non-healthy) and performed classification to detect the tumor in the given samples. Experimental results were more accurate and efficient as compared to the existing tumor detection methods.

A. Markfort *et al.* [25] developed a solution to analyze the medical data over an optimal time period. They defined a correlation between the processing time and the threshold values of pixels to enhance the resolution of input images. Analysis shows that it can process a huge data volume in a minimal time frame and can also calculate the variance of data in the real environment.

MEDICAL IMAGE ANALYSIS

To analyze medical imagery, there are various methods (*i.e.* Classification/Segmentation/Image Encoding/Decoding/Registration/Restoration/ Morphology, *etc.*) Having different outcomes and as per requirement, practitioners can utilize them. For experimental purposes, the Linux platform, python 2.7, and OpenCV 4.x computer vision library were used with the kaggle diabetic retinopathy dataset [32]. Flow chart **1** shows the basic steps to process medical imagery as follows:

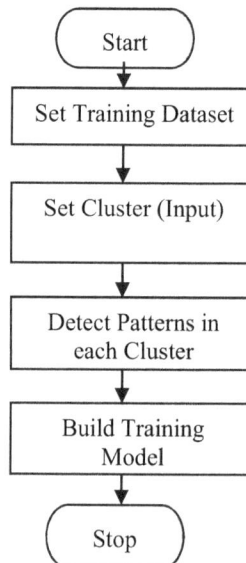

Flowchart (1). Medical image processing.

First of all, input is initialized from a dataset and a method (Classification/Segmentation/Image Encoding/Decoding/Registration/Restoration/Morphology) is selected to generate outcomes as displayed:

Medical Image Classification

Table 1. Medical image classification.

Diabetic retinopathy Level [32]		
Normal	Mild	Moderate

Table 2. Medical image classification.

Diabetic retinopathy Level [32]	
Severe	Proliferative

Tables **1** and **2** show that input image may be classified into different levels with respect to the disease on the basis of features.

Segmentation

Table 3. Image Segmentation

Input Image	Segmentation

Table **3** shows image Segmentation over the input sample. From the output, it can be observed that segmentation can be used to retrieve different regions from the input image for the purpose of analysis.

Image Encoding/Decoding

Table 4. Image Encoding/Decoding

Input Image	Encoded Image
	my\xc9\x02Z#\r7\xa1P`0L\xd9\x8b\xd5\xd3\xf3\xc5\x8e\xae\ xe3\xf0\xc8\x99\x1a\xdb\xce\x0c7\xea\xd8\xddiFmk:M<\xe8\xb4U\xb1-P\xb2\x80> \xda\x99#c\x19\x96#z\xb6\x15\xc3\x02\xda\xb2g\xad\xc5\xd5x\x95\x04\x98\x19\xb5-\x90X\x1e\xe4\ x98Z\xa6-`

Table **4** shows encoding/decoding of input Image. Encoded image can be easily transmitted over network and decoding process is used to obtain the original image.

Registration for Medical Images

Fig. (**3**) shows that using image registration; different features of the input image can be used as references to identify them in the other image.

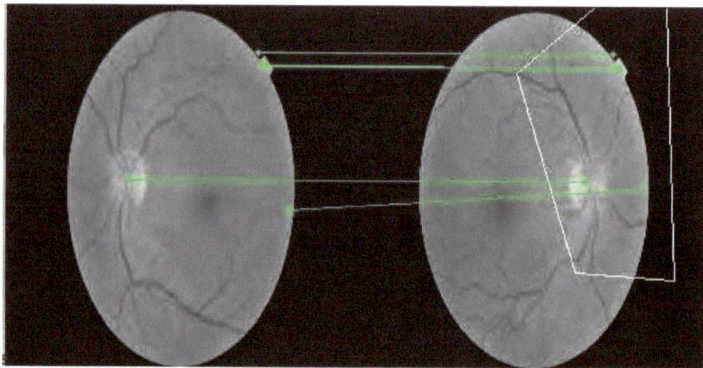

Fig. (3). Medical image registration.

Restoration of Image form Noisy Input

Fig. (**4**) shows the restoration of a refined image from a noisy input image.

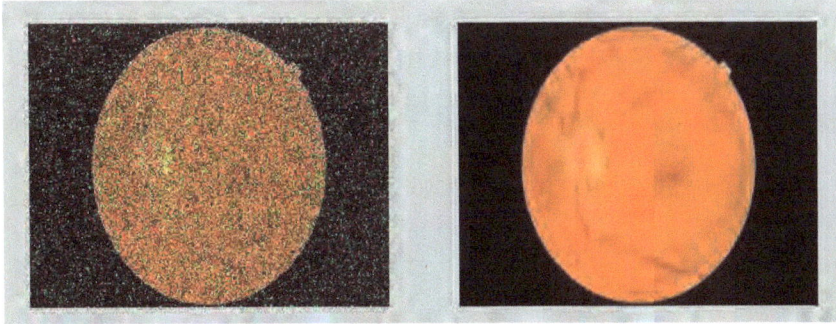

Fig. (4). Medical image restoration.

Morphological Operation for Medical Images

Fig. (**5**) shows a morphological operation over a medical image. It can be observed that its output can be used for disease detection.

Fig. (5). Morphological operation for medical image.

CONCLUSION

In this paper, medical images were analyzed using different methods *i.e.* classification, segmentation, image encoding, decoding, registration, restoration, morphology, *etc.* It can be observed that each method has a different purpose and different applications with respect to the. medial domain. The outcomes of these methods can be used to build training and testing models for prediction purposes. Image classification can be used to determine the type of disease and its level to redirect the diagnosis process, whereas segmentation can be implemented to process different regions of interest in a given input. Encoded images can be transformed into different formats for analysis purposes. The registration process can be used to recognize different parts of an input image; the refined image can be reproduced using the restoration method and morphological operation can be used to process the image at the pixel level. As per the requirements, a particular method may be adapted.

Currently, for analysis, only the diabetic retinopathy dataset was used and in future, an automated disease detection and diagnosis framework will be developed by using these methods for different disease types.

REFERENCES

[1]   X. Chen, X. Wang, K. Zhang, K. M. Fung, T. C. Thai, K. Moore, R. S. Mannel, H. Liu, B. Zheng, and Y. Qiu, "Recent advances and clinical applications of deep learning in medical image analysis", *Medical Image Analysis,* vol. 79, pp. 1-25, 2022.
[http://dx.doi.org/10.1016/j.media.2022.102444]

[2]   F. Liu, J. Tang, J. Ma, C. Wang, Q. Ha, Y. Yu, and Z. Zhou, "The application of artificial intelligence to chest medical image analysis", *Intelligent Medicine,* vol. 1, no. 3, pp. 104-117, 2021.
[http://dx.doi.org/10.1016/j.imed.2021.06.004]

[3]   A. Chopra, and D. C. Verma, "Machine learning-based active contour approach for the recognition of brain tumor progression, book chapter: data science for effective healthcare systems", *Data Science for Effective Healthcare Systems.* 1st Edition, Routledge, CRC Press, Taylor & Francis, pp. 183-198, 2022.

[4]   M. Tsuneki, "Deep learning models in medical image analysis", *Journal of Oral Biosciences,* vol. 64, no. 3, pp. 312-320, 2022.
[http://dx.doi.org/10.1016/j.job.2022.03.003]

[5]   E. G. Mary Kanaga, J. Anitha, and D. Sujitha Juliet, "4D medical image analysis: a systematic study on applications, challenges, and future research directions", *Advanced Machine Vision Paradigms for Medical Image Analysis, Hybrid Computational Intelligence for Pattern Analysis and Understanding,* pp. 97-130, 2021.

[6]   K. Rasheed, A. Qayyum, M. Ghaly, A. A. Fuqaha, A. Razi, and J. Qadir, "Explainable, trustworthy, and ethical machine learning for healthcare: A survey", *Computers in Biology and Medicine,* vol. 149, pp. 1-23, 2022.
[http://dx.doi.org/10.1016/j.compbiomed.2022.106043]

[7]   R. Buettner, M. Bilo, N. Bay, and T. Zubac, "A Systematic Literature Review of Medical Image Analysis Using Deep Learning", *IEEE Symposium on Industrial Electronics & Applications (ISIEA),* pp. 1-4, 2020.
[http://dx.doi.org/10.1109/ISIEA49364.2020.9188131]

[8]   D. Tellez, G. Litjens, J. van der Laak, and F. Ciompi, "Neural Image Compression for Gigapixel Histopathology Image Analysis", *IEEE Transactions on Pattern Analysis and Machine Intelligence,* vol. 43, no. 2, pp. 567-578, 2021.
[http://dx.doi.org/10.1109/TPAMI.2019.2936841]

[9]   P. Seeböck, S.M. Waldstein, S. Klimscha, H. Bogunovic, T. Schlegl, B.S. Gerendas, R. Donner, U. Schmidt-Erfurth, and G. Langs, "Unsupervised Identification of Disease Marker Candidates in Retinal OCT Imaging Data", *IEEE Trans. Med. Imaging,* vol. 38, no. 4, pp. 1037-1047, 2019.
[http://dx.doi.org/10.1109/TMI.2018.2877080] [PMID: 30346281]

[10]   K. Gong, K. Kim, D. Wu, M.K. Kalra, and Q. Li, "Low-dose dual energy CT image reconstruction using non-local deep image prior", *IEEE Nuclear Science Symposium and Medical Imaging Conference (NSS/MIC),* pp. 1-2, 2019.
[http://dx.doi.org/10.1109/NSS/MIC42101.2019.9060001]

[11]   Y. Qi, Y. Guo, and Y. Wang, "Image quality enhancement using a deep neural network for plane wave medical ultrasound imaging", *IEEE Trans. Ultrason. Ferroelectr. Freq. Control,* vol. 68, no. 4, pp. 926-934, 2021.
[http://dx.doi.org/10.1109/TUFFC.2020.3023154] [PMID: 32915734]

[12]   Q. Abbas, M.Q. Yasin, M. Asif, and S. Hussain, "Medical Imaging Analysis using Computer-Assisted Technologies", *Global Conference on Robotics, Artificial Intelligence and Information Technology (GCRAIT),* pp. 202-206, 2022.

[13]   X. Zhou, C. Li, M.M. Rahaman, Y. Yao, S. Ai, C. Sun, Q. Wang, Y. Zhang, M. Li, X. Li, T. Jlang, D. Xue, S. Qi, and Y. Teng, "A comprehensive review for breast histopathology image analysis using classical and deep neural networks", *IEEE Access,* vol. 8, pp. 90931-90956, 2020.
[http://dx.doi.org/10.1109/ACCESS.2020.2993788]

[14]   H. Guan, and M. Liu, "Domain adaptation for medical image analysis: a survey", *IEEE Transactions on Biomedical Engineering,* vol. 69, no. 3, pp. 1173-1185, 2022.
[http://dx.doi.org/10.1109/TBME.2021.3117407]

[15]   X. Wang, H. Chen, C. Gan, H. Lin, Q. Dou, E. Tsougenis, Q. Huang, and M. Cai, "Weakly supervised deep learning for whole slide lung cancer image analysis", *IEEE Transactions on Cybernetics,* vol. 50, no. 9, pp. 3950-3962, 2020.
[http://dx.doi.org/10.1109/TCYB.2019.2935141]

[16]   H. Pinckaers, B. van Ginneken, and G. Litjens, "Streaming convolutional neural networks for end-t--end learning with multi-megapixel images", *IEEE Trans. Pattern Anal. Mach. Intell.,* vol. 44, no. 3, pp. 1581-1590, 2022.
[http://dx.doi.org/10.1109/TPAMI.2020.3019563] [PMID: 32845835]

[17]   B.M. Rashed, and N. Popescu, "Machine learning techniques for medical image processing", *International Conference on e-Health and Bioengineering (EHB),* pp. 1-4, 2021.
[http://dx.doi.org/10.1109/EHB52898.2021.9657673]

[18]   K. Naveen, and R.M.S. Parvathi, "Analysis of Medical Image by using Machine Learning Applications of Convolutional Neural Networks", *2ⁿᵈ International Conference on Artificial Intelligence and Smart Energy (ICAIS),* pp. 115-123, 2022.
[http://dx.doi.org/10.1109/ICAIS53314.2022.9742974]

[19]   S.P. Shayesteh, M. Nazari, A. Salahshour, A.H. Avval, G. Hajianfar, M. Araabi, M. Khateri, H. Abdollahi, H. Arabi, and I. Shiri, "Machine Learning Based Malignancy Prediction in Thyroid Nodules Malignancy: Radiomics Analysis of Ultrasound Images", *IEEE Nuclear Science Symposium and Medical Imaging Conference (NSS/MIC),* pp. 1-2, 2020.
[http://dx.doi.org/10.1109/NSS/MIC42677.2020.9507959]

[20]   T. Zhang, J. Cheng, H. Fu, Z. Gu, K. Zhou, S. Gao, R. Zheng, and J Liu, "Noise adaptation generative adversarial network for medical image analysis", *IEEE Transactions on Medical Imaging,* vol. 39, no. 4, pp. 1149-1159, 2020.
[http://dx.doi.org/10.1109/TMI.2019.2944488]

[21]   J. Geetha, "Thimmiaraja, C. J. Shelke, G. Pavithra, V. K. Sharma, D. Verma, "Deep Learning with Unsupervised and Supervised Approaches in Medical Image Analysis", *2ⁿᵈ International Conference on Advance Computing and Innovative Technologies in Engineering (ICACITE),* pp. 1580-1584, 2022.

[22]   N. Nahar, M.S. Hossain, and K. Andersson, "Medical image analysis using machine learning and deep learning: A comprehensive review", *Rhythms in Healthcare. Studies in Rhythm Engineering,* pp. 147-161, 2022.
[http://dx.doi.org/10.1007/978-981-19-4189-4_10]

[23]   U. Khan, S. Paheding, C. P. Elkin, and V. K. Devabhaktuni, "Trends in deep learning for medical hyperspectral image analysis", *IEEE Access,* vol. 9, pp. 79534-79548, 2021.
[http://dx.doi.org/10.1109/ACCESS.2021.3068392]

[24] A. Sivasangari, and S. Sivakumar, "Helen, S. Deepa, Vignesh, Suja, "Detection of Abnormalities in Brain using Machine Learning in Medical Image Analysis", *International Conference on Sustainable Computing and Data Communication Systems (ICSCDS),* pp. 102-107, 2022.

[25] A. Markfort, A. Baranov, T.M. Conneely, A. Duran, J. Lapington, J. Milnes, A. Mudrov, and I. Tyukin, "Investigating Machine Learning Solutions for High-Speed Data Analysis and Imaging of a Single Photon Counting Detector with Picosecond Timing Resolution", *IEEE Nuclear Science Symposium and Medical Imaging Conference (NSS/MIC),* pp. 1-4, 2021.
[http://dx.doi.org/10.1109/NSS/MIC44867.2021.9875459]

[26] A.S. Panayides, A. Amini, N.D. Filipovic, A. Sharma, S.A. Tsaftaris, A. Young, D. Foran, N. Do, S. Golemati, T. Kurc, K. Huang, K.S. Nikita, B.P. Veasey, M. Zervakis, J.H. Saltz, and C.S. Pattichis, "AI in Medical Imaging Informatics: Current Challenges and Future Directions", *IEEE J. Biomed. Health Inform.,* vol. 24, no. 7, pp. 1837-1857, 2020.
[http://dx.doi.org/10.1109/JBHI.2020.2991043] [PMID: 32609615]

[27] X. Li, W. Shi, Y. Jiao, C. Yang, N. Wang, and Y. Cui, "Medical Ultrasound Image Segmentation Based on Improved MultiResUNet Network", *IEEE International Ultrasonic Symposium (IUS),* pp. 1-3, 2021.
[http://dx.doi.org/10.1109/IUS52206.2021.9593755]

[28] J-H. Huang, T-W. Wu, C-H.H. Yang, and M. Worring, "Deep Context-Encoding Network For Retinal Image Captioning", *IEEE International Conference on Image Processing (ICIP),* pp. 3762-3766, 2021.
[http://dx.doi.org/10.1109/ICIP42928.2021.9506803]

[29] F. Maes, A. Collignon, D. Vandermeulen, G. Marchal, and P. Suetens, "Multimodality image registration by maximization of mutual information", *IEEE Trans. Med. Imaging,* vol. 16, no. 2, pp. 187-198, 1997.
[http://dx.doi.org/10.1109/42.563664] [PMID: 9101328]

[30] L. Chen, P. Bentley, K. Mori, K. Misawa, M. Fujiwara, and D. Rueckert, "Self-supervised learning for medical image analysis using image context restoration", *Medical Image Analysis,* vol. 58, pp. 1-15, 2019.
[http://dx.doi.org/10.1016/j.media.2019.101539]

[31] Z. Ning, S. Zhong, Q. Feng, W. Chen, and Y. Zhang, "SMU-Net: Saliency-Guided Morphology-Aware U-Net for Breast Lesion Segmentation in Ultrasound Image", *IEEE Trans. Med. Imaging,* vol. 41, no. 2, pp. 476-490, 2022.
[http://dx.doi.org/10.1109/TMI.2021.3116087] [PMID: 34582349]

[32] Available from: https://www.kaggle.com/competitions/diabetic-retinopathy-detection/data

CHAPTER 2

IoT Semantic of AI Security Structure for Smart Grid

Ranjit Kumar[1,*], **Rahul Gupta**[1], **Sunil Kumar**[2] and **Neha Gupta**[3]

[1] *Department of Computer Science & Engineering, Maharaja Agrasen University, Baddi, Himachal Pradesh, India*

[2] *Guru Jambheshwar University of Science & Technology, Hisar, Haryana, India*

[3] *Department of Computer Science & Engineering, Institute of Engineering & Technology, Chitkara University, Rajpura, Punjab, India*

Abstract: The integration of the Internet of Things (IoT) and Artificial Intelligence (AI) has revolutionized various industries, and the power sector is no exception. Smart Grid, an advanced power system that employs IoT devices and AI algorithms, promises enhanced efficiency, reliability, and sustainability. However, the proliferation of IoT devices in Smart Grid introduces new security challenges that must be addressed to ensure the integrity and privacy of critical infrastructure. This chapter aims to propose an IoT semantic of AI security structure for the Smart Grid, leveraging advanced AI techniques to detect and mitigate security threats effectively.

Keywords: Anomaly detection, AI security system, IoT, Encryption, Semantic, Smart grid, Threat detection.

INTRODUCTION

The rapid advancement of technology has paved the way for the integration of Smart Grid Internet of Things (IoT) with Smart Home devices, creating a connected ecosystem that offers enhanced energy management, automation, and convenience. This convergence of technologies brings significant benefits, but it also introduces new security challenges and vulnerabilities. Protecting the integrity, confidentiality, and availability of the Smart Grid IoT connected to Smart Home devices is crucial to ensure the efficient and secure operation of these interconnected systems. To address these security concerns, the Semantic AI Security System emerges as a solution designed to enhance the security posture of the Smart Grid IoT connected to Smart Home environments. The Semantic influences the power of artificial intelligence (AI) to detect, mitigate, and respond

---

[*] **Corresponding author Ranjit Kumar:** Department of Computer Science & Engineering, Maharaja Agrasen University, Baddi, Himachal Pradesh, India; E-mail: ranjitpes@gmail.com

Neha Kishore, Pankaj Nanglia, Shilpa Gupta & Ashutosh Kumar Dubey (Eds.)

to potential threats in real-time, thereby safeguarding the integrity and confidentiality of the system.

The objective of this chapter is to provide a comprehensive overview of the Semantic AI Security System in the context of the Smart Grid IoT connected to Smart Home devices. This introduction sets the stage by highlighting the importance of security in this interconnected ecosystem and providing a brief overview of the semantic. Importance of Security in Smart Grid IoT and Smart Home: The integration of Smart Grid IoT with Smart Home devices has revolutionized the way we manage and consume energy. However, this interconnected environment also exposes vulnerabilities that can be exploited by malicious actors. Unauthorized access, data breaches, and manipulation of energy consumption patterns are some of the security concerns that need to be addressed. Failure to ensure robust security measures can lead to significant financial losses, privacy breaches, and disruption of critical services.

Overview of the Semantic: The Semantic AI Security System is specifically designed to mitigate the security risks faced by the Smart Grid IoT connected to Smart Home devices. It combines advanced AI algorithms, machine learning techniques, and security protocols to detect, analyze, and respond to potential threats. By continuously monitoring the network traffic, device behavior, and communication patterns, the Semantic can identify anomalies and take appropriate actions to mitigate the risks.

The Semantic encompasses several key components, including threat detection mechanisms, anomaly detection techniques, authentication and access control, encryption and data privacy, incident response and recovery, system updates and patch management, as well as user awareness and training. These components work together to create a robust security framework that ensures the integrity and confidentiality of the interconnected system. The primary objective of this chapter is to provide a comprehensive understanding of the Semantic AI Security System in the context of the Smart Grid IoT connected to Smart Home devices. The chapter aims to:

- Explore the security challenges and vulnerabilities associated with this interconnected ecosystem.
- Explain the key components and features of the Semantic AI Security System and how they address the security concerns.
- Present real-world case studies and examples to demonstrate the effectiveness of the Semantic in mitigating security threats.
- Evaluate the performance of the Semantic through experimental scenarios and metrics.

• Discuss future research directions and emerging trends in AI-driven security systems for Smart Grid IoT and Smart Home applications.

By achieving these objectives, this chapter aims to contribute to the existing body of knowledge in the field of Smart Grid IoT connected to home security and provide valuable insights into the Semantic AI Security System as a robust solution for threat mitigation.

IOT ENABLED SMART GRID

IoT is the future of grid networks. IoT for the smart grid is defined by the NIST as fusing the current emerging ICT grid with the old [1]. The smart grid, in contrast to conventional power networks, can maintain or regulate the demand for power dis- tribution, accomplish efficient power delivery, and reduce energy losses [2]. Its capacity to adapt to fluctuating supply and demand is what makes it "smart". The "smart" energy grid of today is made possible by technologies [3].

Outline of Smart Grid Design

IoT enabled smart grid is an advanced and interconnected infrastructure that combines traditional power systems with digital technologies and communication networks. It brings together devices, sensors, meters, and control systems to enable more efficient, reliable, and sustainable energy management. The integration of IoT technologies into the traditional power grid transforms it into a smart and dynamic ecosystem capable of real-time monitoring, analysis, and control.

Key Components of Smart Grid IoT

Smart Meters

Smart meters are digital devices installed at consumer premises that enable two-way communication between the utility provider and the consumer. They provide real- time energy consumption data, allowing consumers to monitor and manage their energy usage efficiently.

Advanced Metering Infrastructure (AMI)

AMI refers to the network infrastructure that supports the communication between smart meters and utility providers. It enables automated meter reading, remote monitoring, and control of energy consumption.

Distribution Automation

Distribution automation involves the deployment of sensors, control devices, and communication networks throughout the distribution system. It enables real-time monitoring, fault detection, and self-healing capabilities to optimize power distribution and reduce outage durations.

Demand Response Systems

Demand response systems allow utility providers to adjust energy consumption during peak demand periods. Through IoT connectivity, consumers can participate in demand response programs, allowing them to reduce energy usage during peak times and receive incentives in return.

Energy Management Systems

Energy management systems leverage IoT connectivity and data analytics to provide consumers with detailed insights into their energy usage patterns. These systems allow consumers to optimize their energy consumption, track costs, and implement energy-efficient practices.

Benefits of Smart Grid IoT

Improved Efficiency

Smart Grid IoT enables real-time monitoring and control of energy consumption, allowing utility providers and consumers to optimize energy usage and reduce wastage. It enables dynamic load management, efficient distribution, and improved demand response capabilities.

Enhanced Reliability

IoT connectivity in the Smart Grid enables proactive monitoring, fault detection, and self-healing capabilities. This results in quicker response times, reduced downtime, and improved reliability of power supply.

Integration of Renewable Energy

Smart Grid IoT facilitates the seamless integration of renewable energy sources into the power grid. It enables better management of distributed energy resources, grid stability, and the optimization of renewable energy generation.

Consumer Empowerment

With Smart Grid IoT, consumers have access to real-time energy consumption data, allowing them to make informed decisions about their energy usage. They can actively participate in energy management, implement energy-saving practices, and reduce costs.

Environmental Sustainability

By optimizing energy consumption, integrating renewable energy sources, and reducing wastage, Smart Grid IoT contributes to a more sustainable and eco-friendly energy ecosystem. It enables the reduction of greenhouse gas emissions and promotes the adoption of clean energy technologies.

Challenges and Considerations

While Smart Grid IoT offers numerous benefits, it also poses challenges and considerations:

Security and Privacy

The interconnected nature of Smart Grid IoT introduces security vulnerabilities and privacy concerns. Protecting data integrity, ensuring secure communication, and safeguarding against cyber threats are crucial considerations.

Interoperability

The integration of diverse devices, protocols, and systems in Smart Grid IoT requires seamless interoperability to ensure efficient communication and data exchange.

Scalability

As the data volumes with devices increase, scaling up the infrastructure and ensuring smooth operation become essential challenges.

Regulatory and Policy Frameworks

The implementation of Smart Grid IoT requires the development of appropriate regulatory and policy frameworks to address issues such as data privacy.

Various smart devices can be connected *via* smart grid networks [4]. In order to guarantee that grid electricity is controlled securely, it may also reduce human meddling by monitoring meters, home portals, and other important equipment [5]. The NIST specified smart grid model as depicted in Fig. (1) is what most grids adhere to. NIST categorizes smart grid entities into four main categories: service providers, operations, distributors, and marketplaces. The home users make up the three (3) primary user categories. The smart grid may share data in both directions with all of its components by using IoT [6]. Given the utilisation of smart meters and other intelligent devices on the end-customer side, in addition to sensors, drives, and other intelligent devices, the touch is plausible. This enables the monitoring of energy demand and consumption while allowing users to observe and alter their behaviour [8].

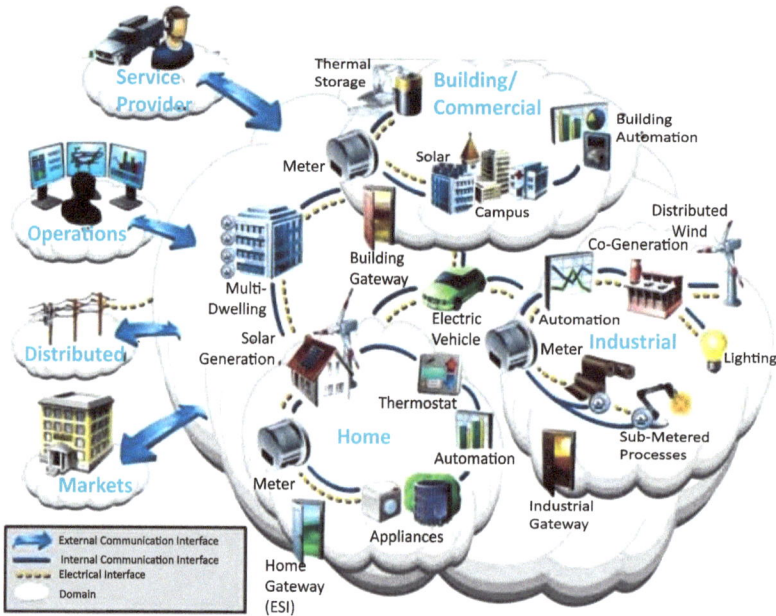

Fig. (1). NIST customer field smart grid model [7].

The introduction of IoT to the smart grid has a lot of benefits. For instance, the IoT smart grid makes it simple to access the Advanced Metering Infrastructure (AMI) [9]. Real-time computation accessibility offers consumers and manufacturers crucial signals to more effectively meet their energy consumption and supply requirements [10].

In case of any aggravation or risk, whether internal or external, the IoT smart grid can immediately repair itself (self-recovery) [11]. Additionally, it enables the self-rebuilding of the system through intricate reconfigurations in order to regain electricity following threats, catastrophic occurrences, power outages, or breakdown of the network components. In addition, it can identify the source of energy leakage while building a micro-scale grid and self-sufficiently protected islands in the case of a power outage [12]. This improves the modelling, analysis, and performance of the power grid, enhancing the consistency of the power grid. This can easily track demand and request reactions on the smart grid using the IoT smart grid, enabling for the effect of complicated power price components [13]. By enforcing higher costs during peak times to deter consumption and lower prices during off-peak times to promote less utilisation and idle usage of power, competitive energy pricing improves peak load management capabilities [14].

Real-time, quick, and bidirectional information exchange enables better communication with energy end users. It gives utilities crucial information about client usage patterns that improves the linked grid [15]. The massive sensor installation, together with data processing and real-time networking tools, aids in determining the positions of individual grid pieces [16]. This contributes to a reliable transfer of power by helping to manage energy, prepare for present and future prospects, and make the transmission lines and transformers perform in this way. These evaluations may enable the steadfast efficiency of the gearbox structure to be increased by clearly recognising the indications of line problems, reducing the probability of catastrophic failure, and lowering operating and maintenance expenses in these lines [17]. Fig. (**2**) illustrates the attacker trying to connect *via* different IoT connected in smart home to get the smart grid service.

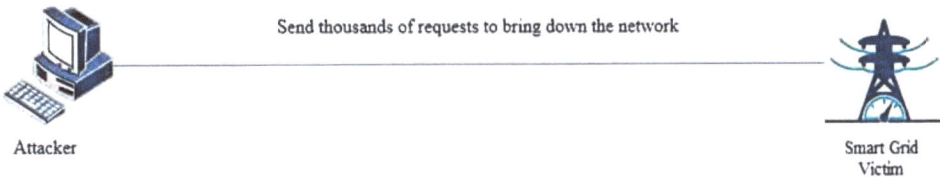

Fig. (2). Attacker request to down smart grid network connected to smart home.

In a Smart Grid IoT connected home environment, potential attackers can exploit vulnerabilities to compromise the security and privacy of the system. Here are some examples of attackers in the Smart Grid IoT connected home context:

1. **Malicious External Hackers:** These attackers are external entities with the intent to breach the Smart Grid IoT infrastructure and gain unauthorized access to sensitive information or disrupt services. They may target vulnerabilities in communication protocols, weak authentication mechanisms, or unpatched software in smart home devices or the Smart Grid infrastructure.

2. **Botnets and IoT-based Attacks:** Botnets are networks of compromised IoT devices controlled by a remote attacker. These devices can be used to launch distributed denial-of-service (DDoS) attacks, overwhelm network resources, or gain unauthorized access to the Smart Grid IoT infrastructure. By compromising

vulnerable smart home devices, attackers can build powerful botnets to launch attacks.

Smart Grid Features

The existing grid is currently facing everyday difficulties, like interruption in a specific location. Though, this can only be established if the end user reports the incident. Instead, by using smart grid monitoring, we can proactively respond to the outages. The secret to simple integration and consumer dependability is smart grid technology [18]. With a smaller workforce and a focus on providing all users with reliable, secure, and sustainable energy, this system can swiftly address issues in the present system. The following are advantages of smart grid technology, per [19, 20 - 22].

Smarter Use of Energy

Overall, the smart grid enables a more intelligent and efficient use of energy by leveraging real-time data, automation, and advanced technologies. It empowers consumers to make informed decisions about their energy consumption, optimizes grid operations, facilitates the integration of renewable energy sources, and supports the transition towards a more sustainable and resilient energy system [21].

Cleaner Use of Energy

Smart grid solutions are intended to minimise the maximum load on distribution feeders, use less batteries, and are carbon-efficient. In order to provide sustainable solutions, the use of integrated micro-grid machineries, it is possible to support the whole distribution network [23].

Lesser Costs

It is important to note that while the implementation of a smart grid involves upfront investments in infrastructure and technologies, the long-term cost savings and operational efficiencies often outweigh the initial costs, making it a financially beneficial solution for utilities and consumers alike [24].

Enhanced Transportation and Parking

Smart IoT sensors may gather data in real-time to provide information to drivers and authorities. In the end, this would shorten commute times, improve parking alternatives, notify drivers of traffic incidents and townscape structural damage, and enable electronic payment at toll booths and parking meters. Wireless electric car charging is another projected benefit of IoT technology in the future [25].

Backing in Waste Managing

Smart cities will gain from the smart network by increasing yield and lowering the cost of their waste management solutions. IoT apps may offer real-time data to track inventories and reduce fraud. Increased time and scheduling for truck routes may result from cloud-based traffic management and monitoring. Smart energy analytics may gather information on water flow, pressure, temperature, and other things, enabling users to monitor their usage patterns [26].

Energy Enablement

Energy enablement in emerging countries refers to the process of providing access to reliable and affordable energy services to support social and economic development. Access to energy is crucial for improving living standards, promoting economic growth, and addressing various development challenges. This may provide up opportunities for economic growth.

Through a flexible communication network that provides greater synchronisation among the IoT enables grid means. The grid will be able to fulfil both current and future energy demands thanks to this, which has been shown to increase production and accuracy. But there are always difficulties with technology growth, particularly in terms of security. The challenges with IoT security in the context of smart grids are covered in the section that follows.

SMART GRID SECURITY ISSUES RELATED WORKS ON IOT

Numerous research studies have been conducted to address smart grid security issues related to IoT. Here are some notable works in this area:

K. N. Cheung *et al*. (2014) discusses security challenges in the smart grid and IoT, focusing on authentication, data integrity, privacy, and network resilience [27]. It provides an overview of potential threats and proposes security mechanisms and protocols to mitigate them. F. L. Lewis *et al*. (2015) presents an overview of security challenges in the smart grid and proposes a distributed multi-agent framework for enhancing grid security. It discusses the integration of IoT devices and suggests security measures for protecting critical infrastructure [28]. M. M. A. Alqerm *et al*. (2016) provides an extensive analysis of security issues in smart grid communication networks, with a focus on IoT technologies [29]. It covers authentication, access control, data integrity, privacy, and intrusion detection, and presents various security solutions proposed in the literature. M. S. Hossain *et al*. (2018) explores security and privacy challenges in smart grids, including IoT-related concerns. It discusses potential threats, such as network attacks, data tampering, and privacy breaches, and proposes secure communication protocols, cryptographic techniques, and intrusion detection systems to address these issues [30].

A. L. Munoz *et al.* (2020) focuses on secure communication in smart grids with an emphasis on IoT devices. It reviews various communication protocols and technologies for IoT deployment in smart grids, identifies potential security vulnerabilities, and proposes security measures for ensuring confidentiality, integrity, and availability of grid communications [31]. S. Misra *et al.* (2016) focusing on communication aspects, this review paper examines security issues in Smart Grid communications. It discusses communication protocols, vulnerabilities, and attacks in the context of Advanced Metering Infrastructure (AMI), Home Area Networks (HANs), and Wide Area Networks (WANs). The authors review security solutions such as encryption, authentication, and intrusion detection. The paper highlights the importance of secure communication protocols, network segmentation, and secure key management in Smart Grid communications [32].

D. K. Mishra *et al.* (2018) presents a comprehensive analysis of security threats and vulnerabilities in Smart Grid deployments. It categorizes the threats into physical, cyber, and social aspects. The authors review existing security solutions, including encryption, authentication, intrusion detection, and anomaly detection. They also highlight emerging technologies such as blockchain and machine learning for Smart Grid security. The paper provides valuable insights into the current state of Smart Grid security and proposes future research directions [33].

These works highlight the importance of addressing security challenges in smart grid systems, particularly in the context of IoT integration. They provide insights into potential threats, propose security mechanisms and protocols, and suggest ways to protect smart grid infrastructure, data, and communication networks from unauthorized access, tampering, and attacks.

The necessity for a thorough assessment of smart grid security vulnerabilities has been raised in the literature. The energy sector is crucial, and there is little doubt that IoT-enabled smart grids will encounter cybersecurity concerns. They are especially susceptible to threats and cover many locations. Effective information sharing over the IoT is significantly hampered by the malware threat. A cyberattack on these IoT devices might result in the loss of important data or possibly a stoppage in business operations. For instance, in 2015, a cyberattack on the Ukrainian power grid gave hackers access to the SCADA network and allowed them to monitor and halt the network using the BlackEnergy virus. Over 700,000 consumers lost access to electricity as a result, resulting in a significant blackout.

The large number of methods in use that don't improve device security makes IoT protection more challenging. Unlike desktops, we would be able to reinstall or wipe out the entire network, but the majority of IoT devices do not yet support that. According to a study by Salameh, Dhainat, and Benkhelifa, these sensor efficacies vary because many IoT devices are mostly dependent on their manufacturers. Users are unable to tamper with the devices because the creators have locked them. In this article, IoT security concerns are discussed in relation to the smart grid and are divided into three categories: component security concerns, system security concerns, and network security concerns. These studies offer insightful information on the security concerns and difficulties associated with deploying the Smart Grid to smart homes. They address a wide range of topics, including threats, vulnerabilities, security mechanisms, privacy concerns, and emerging technologies. By reviewing existing research, these chapters contribute to the understanding of Smart Grid security and provide guidance for developing robust security frameworks in Smart Grid deployments to smart home.

In Table **1**, the potential attack kinds and their categorization related to the IoT deployment in SG are summarised.

Table 1. IoT enabled smart grid security concern and threat kinds.

Security Issue Cataloging	Threat Type	Study
Module Security	• MITM • Misleading data from smart metres • Phasor Data Connector (PDC) and • Phasor Measurement Unit (PMU) threats	[8, 14, 17]
System Security	• False information threats, • Threats on the Delivery Management System (DMS), • Threats on the Energy Management System (EMS), • And man-in-the-middle attacks • Stealth-listening • WAMPAC system • DOS threat • Malware threat	[5, 23, 27, 28]
Network Security	• DoS threat • Routing threat • WSN threat	[3, 15, 29]

IoT has mostly focused on the demand side. The mitigating strategy for the raised security risks is described in the next section.

THREAT MITIGATION

Threat mitigation in a smart grid involves the implementation of measures and technologies to protect the grid from potential risks and ensure its reliable and secure operation. Here are some key areas and strategies for threat mitigation in a smart grid.

Cybersecurity

With increased connectivity and digitalization, smart grids face a higher risk of cyber threats. To mitigate these threats, several measures can be implemented, including:

- Network Segmentation: Dividing the smart grid network into segments and implementing firewalls to restrict unauthorized access and limit the spread of cyberattacks.
- Strong Authentication: Implementing multi-factor authentication mechanisms to ensure that only authorized personnel can access critical grid components.
- Encryption: Using robust encryption techniques to protect data transmitted over the network and prevent unauthorized access or data manipulation.
- Intrusion Detection and Prevention Systems (IDPS): Deploying IDPS solutions that monitor network traffic, detect potential threats or attacks, and take preventive measures.
- Regular Patching and Updates: Keeping all software and firmware up to date with the latest security patches to address vulnerabilities and protect against known threats.

Physical Security

Protecting physical infrastructure is crucial for maintaining the integrity and reliability of the smart grid. Some key physical security measures include:

- Restricted Access: Implementing access controls, surveillance systems, and secure perimeters to prevent unauthorized physical access to critical grid components.
- Backup Power and Redundancy: Ensuring backup power supplies and redundancy in critical components to mitigate the impact of physical attacks or natural disasters.
- Monitoring Systems: Deploying sensors, cameras, and alarm systems to detect and respond to physical threats such as tampering, vandalism, or unauthorized activities in substations or power generation facilities.

• Security Personnel and Training: Employing trained security personnel and providing them with proper training to respond effectively to physical security incidents.

Resilience and Disaster Recovery

Smart grids should have robust resilience and disaster recovery plans to minimize the impact of potential threats, including natural disasters or equipment failures. Some strategies for resilience and disaster recovery include:

• Redundancy and Diversity: Implementing redundant systems and diverse energy sources to ensure continuity of power supply in the event of disruptions.
• Grid Monitoring and Rapid Response: Deploying real-time monitoring systems to detect anomalies, failures, or cyberattacks and enabling quick response and recovery actions.
• Backup Systems and Emergency Preparedness: Having backup systems, emergency generators, and contingency plans to restore critical services during disruptions.
• Regular Testing and Drills: Conducting periodic testing and drills to validate the effectiveness of resilience and disaster recovery plans and identify areas for improvement.

Data Privacy

Smart grids generate a vast amount of data, including energy consumption patterns and customer information. Protecting the privacy of this data is crucial. Mitigation measures include:

• Data Encryption and Anonymization: Encrypting sensitive data and anonymizing it whenever possible to protect individual privacy.
• Access Controls and Data Governance: Implementing strict access controls and data governance policies to ensure that only authorized personnel can access and handle sensitive data.
• Compliance with Data Protection Regulations: Adhering to applicable data protection regulations and industry best practices to safeguard customer data and privacy.
• Regular Auditing and Monitoring: Conducting regular audits and monitoring activities to identify any data privacy risks or breaches and take appropriate corrective actions.

These are some key aspects of threat mitigation in a smart grid. Implementing a holistic approach that combines cybersecurity, physical security, resilience, and privacy measures are crucial to ensuring the reliable and secure operation of the smart grid infrastructure. To incorporate defects and software upgrades equally and effectively, IoT systems must be scalable [34]. Unfortunately, the majority of programmers today create software without thinking creatively about prospective firmware updates. If firmware changes are not intended to be available, deployment might be challenging. Given the enormous size of a smart IoT system, periodic firmware upgrades are the practical and reasonable course of action as opposed to the comprehensive replacement of outmoded systems within reach. Regardless of the level of network security, when firms combine new and outdated technology, cybersecurity issues are dramatically increased.

All grid systems must be stable in their relationship to the environment. To prevent unauthorised physical entrance. Data such as authorisation, identification, use, and account details may be kept in hacked systems as a result of remote exposure by unauthorised personnel. Because hackers can utilise remote wiping technologies for nefarious purposes [35]. This creates a focal point from which people looking to disrupt the network, including hackers and angry ex-employees, may simply command the whole Internet of Things smart grid.

Always be crucial when building IoT for smart grid applications since the IoT devices might be utilised for law enforcement or surveillance. However, it also has the potential to be a double-edged weapon that a terrorist might use. As a result, from the start, both makers and users should make sure that the smart grid application is free of malicious code and bypasses. In order to stop DoS attacks, a workable network layer programming method is needed [35]. Utilising the quick hopping Internet Protocol (IP) can fix this. It gives customers a simple method to control the subject matter and web address of their contact sessions.

PROPOSED IOT PARTS TO SECURE SMART GRID

The proposed additional model will be helpful to assure security observance owing to unforeseen cybersecurity dangers, notably on IoT. Fig. (3) displays the mentioned recommended model.

AI Access Control

Artificial Intelligence Access control refers to the mechanisms and policies used to regulate and manage user access to resources in a computer system or network. It ensures that only authorized individuals or entities can access specific information or perform certain actions. Access control mechanisms include authentication (verifying the identity of users), authorization (determining what resources users can access), and auditing (keeping records of user activities).

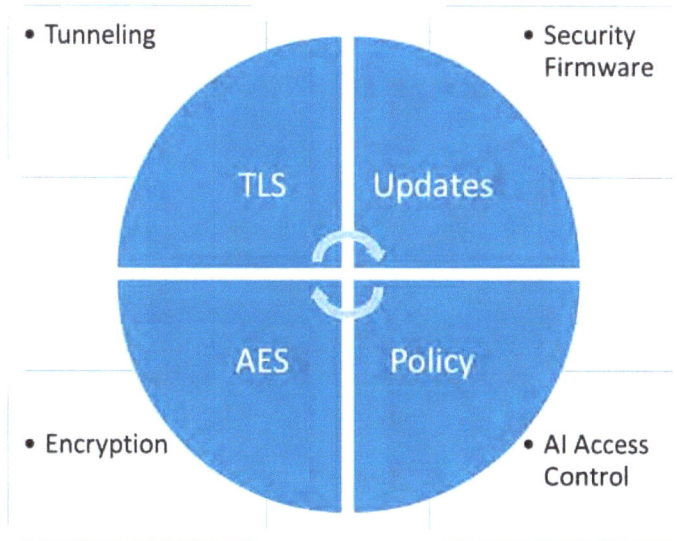

Fig. (3). IoT smart grid security consolidation model.

Here's how AI can be utilized for access control in the smart grid:

Identity Verification

AI can be used for identity verification of individuals or devices seeking access to the smart grid. Facial recognition, fingerprint scanning, voice recognition, or other biometric authentication methods can be integrated with AI algorithms to ensure that only authorized personnel or devices are granted access.

Anomaly Detection

AI-powered anomaly detection algorithms can monitor and analyze user behavior patterns within the smart grid. By establishing baseline behavior profiles, AI systems can identify and flag any suspicious activities that deviate from normal patterns. This helps in detecting unauthorized access attempts, potential cyber threats, or anomalous behavior that may indicate system vulnerabilities.

Real-time Monitoring

AI algorithms can continuously monitor access activities in real-time. By analyzing data from access logs, user authentication, and system events, AI systems can detect and respond to any unauthorized access attempts promptly. Real-time monitoring allows for immediate identification and mitigation of security breaches, reducing the risk of potential damage or disruptions.

Adaptive Access Control

AI can enhance access control mechanisms by dynamically adapting access privileges based on user behavior and contextual factors. AI systems can analyze historical data and real-time inputs to make intelligent decisions about granting or revoking access rights. This adaptive approach ensures that access permissions align with changing circumstances and potential risks.

Threat Intelligence and Predictive Analytics

AI can leverage advanced threat intelligence and predictive analytics to identify emerging security threats and vulnerabilities in the smart grid. By analyzing large volumes of data from diverse sources, AI systems can detect patterns, correlations, and indicators of potential cyberattacks or unauthorized access attempts. This enables utilities to proactively strengthen their access control measures and address vulnerabilities before they are exploited.

User Access Policy Management

AI systems can assist in managing user access policies within the smart grid. By analyzing user roles, privileges, and access history, AI algorithms can recommend and enforce access control policies based on predefined rules and regulations. This helps utilities ensure that access permissions are aligned with the principle of least privilege, where users are granted only the necessary access rights to perform their specific tasks.

Continuous Authentication

AI-powered continuous authentication systems can dynamically monitor and verify user identities throughout their interaction with the smart grid. This involves the use of AI algorithms to analyze multiple factors such as behavioral biometrics (keystroke dynamics, mouse movement), location data, and device characteristics to continuously assess the legitimacy of user access. Continuous authentication enhances security by mitigating the risks associated with stolen credentials or compromised accounts.

Security Patching

Security patching involves the process of applying updates or fixes to software systems to address identified security vulnerabilities or bugs. Software vulnerabilities can be exploited by attackers to gain unauthorized access, cause damage, or compromise the system's integrity. Regularly installing security patches and updates helps protect against known vulnerabilities and strengthens the overall security posture.

Tunneling

Tunneling is a technique commonly employed in Smart Grid IoT security to ensure secure and private communication between devices and networks. It involves encapsulating data packets from one network protocol within the packets of another protocol, thereby creating a virtual tunnel through which the data can traverse securely. In the context of Smart Grid IoT, tunneling plays a crucial role in protecting sensitive information and ensuring the integrity of data transmission.

Here are some key aspects of tunneling in Smart Grid IoT security:

- Secure Data Transmission: Tunneling enables the secure transmission of data between different components of the Smart Grid IoT infrastructure. It ensures that data exchanged between devices, sensors, gateways, and control systems remain confidential and protected from unauthorized access or eavesdropping.
- Network Segmentation: Tunneling allows for the segmentation of the Smart Grid IoT network into virtual private networks (VPNs) or separate network segments. Each segment can have its own security policies and access controls, preventing unauthorized access to critical infrastructure and reducing the attack surface.
- Interoperability: Tunneling protocols facilitate interoperability by encapsulating different network protocols within a standardized protocol. This allows devices and systems that use different protocols to communicate securely, enabling seamless integration of diverse components in the Smart Grid IoT ecosystem.
- VPN Technologies: Virtual Private Network (VPN) technologies, such as IPsec (Internet Protocol Security) and SSL/TLS (Secure Sockets Layer/Transport Layer Security), are commonly used for tunneling in Smart Grid IoT security. VPNs create secure tunnels over public or untrusted networks, providing a secure communication channel for sensitive data.
- Secure Remote Access: Tunneling allows for secure remote access to Smart Grid IoT devices and systems. It enables authorized personnel to remotely connect to and manage devices within the network without compromising security. Through tunneling, remote access can be encrypted and authenticated, preventing unauthorized access to critical infrastructure.
- Defense Against Network Attacks: Tunneling can provide a defense against certain network-based attacks by encapsulating and encrypting data packets. This protects the data from tampering, spoofing, or interception, reducing the risk of data manipulation or unauthorized access.
- VPN Concentrators and Gateways: VPN concentrators and gateways are deployed in the Smart Grid IoT infrastructure to manage and secure the tunneling process. These devices establish and manage tunnels, enforce security

policies, and authenticate connections, ensuring secure communication between devices and networks.

Encryption

Encryption is the process of converting plaintext (readable data) into ciphertext (encrypted data) using an encryption algorithm and a cryptographic key. It ensures that data remains confidential and secure even if it's intercepted or accessed by unauthorized parties. Encryption is widely used to protect sensitive information during storage, transmission, or communication. Only authorized individuals possessing the decryption key can decipher and access the original data [37].

Overall, these security practices play crucial roles in safeguarding computer systems, networks, and sensitive information from unauthorized access, data breaches, and other security threats. Organizations and individuals should adopt and implement these measures to mitigate risks and enhance the security of their digital assets [38].

CONCLUSION AND FUTURE WORK

In this chapter, we emphasized that the management activities and sensitive data transmitted make it imperative to secure the grid. It highlights the significance of the proposed IoT semantic of AI security structure for Smart Grid. It emphasizes the importance of robust security measures in ensuring the trustworthiness and resilience of IoT-based Smart Grids. In conclusion, IoT-enabled smart grid security is a crucial aspect of ensuring the reliable and secure operation of modern power systems. With the increasing adoption of technologies in the energy sector, the potential vulnerabilities and security threats have also multiplied. However, significant progress has been made in addressing these challenges and developing robust security measures for IoT-enabled smart grids. One of the key findings is that a multi-layered security approach is necessary to protect IoT-enabled smart grids. This approach involves securing the physical infrastructure, communication networks, and the data transmitted between various devices and systems. Encryption, authentication, access control, and intrusion detection systems play vital roles in ensuring the confidentiality, integrity, and availability of smart grid data and operations.

Additionally, collaboration among stakeholders is crucial in enhancing the security of IoT-enabled smart grids. Utilities, device manufacturers, regulators, and researchers need to work together to develop and implement standards, best practices, and guidelines for secure IoT deployments in the energy sector. Regular security audits, vulnerability assessments, and penetration testing should be conducted to identify and mitigate potential weaknesses in the system.

While significant progress has been made in IoT-enabled smart grid security, there are still several areas that require further research and development. Some potential areas for future work include:

1. Threat intelligence and analytics: Developing advanced analytics and machine learning algorithms to identify and detect emerging security threats in real-time. This can help in proactive threat mitigation and faster incident response.

2. Privacy preservation: Addressing the privacy concerns associated with the collection and analysis of massive amounts of data in smart grids. Exploring techniques such as differential privacy to ensure the anonymity and confidentiality of sensitive information.

a. Blockchain technology: Investigating the use of blockchain for enhancing the security and trustworthiness of IoT-enabled smart grids. Blockchain can provide a decentralized and immutable ledger for secure transactional and data management.

b. Resilience and disaster recovery: Designing resilient architectures and mechanisms to ensure the continuity of smart grid operations during natural disasters, cyber-attacks, or system failures.

c. Standardization and regulation: Working towards the development of global standards and regulations specifically tailored for IoT-enabled smart grid security. This can help in achieving interoperability, consistency, and compliance across different regions and stakeholders.

By focusing on these areas and continuing to make innovation in IoT-enabled smart grid security, we can create a more secure, reliable, and efficient energy infrastructure for the future.

REFERENCES

[1]   K. Kandasamy, S. Srinivas, K. Achuthan, and V.P. Rangan, "IoT cyber risk: a holistic analysis of cyber risk assessment frameworks, risk vectors, and risk ranking process", *EURASIP Journal on Information Security,* vol. 2020, no. 1, p. 8, 2020.
       [http://dx.doi.org/10.1186/s13635-020-00111-0]

[2]   P. Sethi, and S.R. Sarangi, "Internet of Things: Architectures, Protocols, and Applications", *J. Electr. Comput. Eng.,* vol. 2017, pp. 1-25, 2017.
       [http://dx.doi.org/10.1155/2017/9324035]

[3]   B Champaty, SK Nayak, G Thakur, and B Mohapatra, "Development of Bluetooth, Xbee and Wi-Fi-based Wireless Control Systems for Controlling Electric- Powered Robotic Vehicle Wheelchair Prototype. Robotic Systems: Concepts, Methodologies", *Tools and Applications,* pp. 1048-1079, 2020.

[4]   V. Hassija, V. Chamola, V. Saxena, D. Jain, P. Goyal, and B. Sikdar, "A Survey on IoT Security: Application Areas, Security Threats, and Solution Architectures", *IEEE Access,* vol. 7, pp. 82721-82743, 2019.
[http://dx.doi.org/10.1109/ACCESS.2019.2924045]

[5]   S.K. Rathor, and D. Saxena, "Energy management system for smart grid: An overview and key issues", *Int. J. Energy Res.,* vol. 44, no. 6, pp. 4067-4109, 2020.
[http://dx.doi.org/10.1002/er.4883]

[6]   X. Fang, S. Misra, G. Xue, and D. Yang, "Smart Grid — The New and Improved Power Grid: A Survey", *IEEE Commun. Surv. Tutor.,* vol. 14, no. 4, pp. 944-980, 2012.
[http://dx.doi.org/10.1109/SURV.2011.101911.00087]

[7]   S.K. Goudos, P. Sarigiannidis, P.I. Dallas, and S. Kyriazakos, Springer; 2019.

[8]   F. Dalipi, and S.Y. Yayilgan, "Security and Privacy Considerations for IoT Application on Smart Grids: Survey and Research Challenges", *Proc. of the 4th IEEE International Conference on Future Internet of Things and Cloud Workshops (FiCloudW),* pp. 63-68, 2016.

[9]   B. Jelacic, I. Lendak, S. Stoja, M. Stanojevic, and D. Rosic, "Security Risk Assessment-based Cloud Migration Methodology for Smart Grid OT Services", *Acta Polytech. Hung.,* vol. 17, no. 5, pp. 113-134, 2020.
[http://dx.doi.org/10.12700/APH.17.5.2020.5.6]

[10]   M.Z. Gunduz, and R. Das, "Cyber-security on smart grid: Threats and potential solutions", *Comput. Netw.,* vol. 169, pp. 107094-107094, 2020.
[http://dx.doi.org/10.1016/j.comnet.2019.107094]

[11]   L. Tightiz, and H. Yang, "A Comprehensive Review on IoT Protocols' Features in Smart Grid Communication", *Energies,* vol. 13, no. 11, pp. 2762-2762, 2020.
[http://dx.doi.org/10.3390/en13112762]

[12]   A. Ghasempour, "Internet of Things in Smart Grid: Architecture, Applications, Services, Key Technologies, and Challenges", *Inventions (Basel),* vol. 4, no. 1, pp. 22-22, 2019.
[http://dx.doi.org/10.3390/inventions4010022]

[13]   M. Faheem, S.B.H. Shah, R.A. Butt, B. Raza, M. Anwar, M.W. Ashraf, M.A. Ngadi, and V.C. Gungor, "Smart grid communication and information technologies in the perspective of Industry 4.0: Opportunities and challenges", *Comput. Sci. Rev.,* vol. 30, pp. 1-30, 2018.
[http://dx.doi.org/10.1016/j.cosrev.2018.08.001]

[14]   S. Eom, and J.H. Huh, "The opening capability for security against privacy infringements in the smart grid environment", *Mathematics,* vol. 6, no. 10, pp. 202-202, 2018.
[http://dx.doi.org/10.3390/math6100202]

[15]   NIST Office of the National Coordinator for Smart Grid Interoperability (2013), "NIST Framework and Roadmap for Smart Grid Interoperability Standards, Release 3.0:, Available at: http://www.nist.gov/smartgrid/upload/Draft-NIST-SG-Framework-3.pdf

[16]   T. Alladi, V. Chamola, and S. Zeadally, "Industrial Control Systems: Cyberattack trends and countermeasures", *Comput. Commun.,* vol. 155, pp. 1-8, 2020.
[http://dx.doi.org/10.1016/j.comcom.2020.03.007]

[17]   K. Kimani, V. Oduol, and K. Langat, "Cyber security challenges for IoT-based smart grid networks", *Int. J. Crit. Infrastruct. Prot.,* vol. 25, pp. 36-49, 2019.
[http://dx.doi.org/10.1016/j.ijcip.2019.01.001]

[18]   A. Ghosal, and M. Conti, "Key management systems for smart grid advanced metering infrastructure: A survey", *IEEE Commun. Surv. Tutor.,* vol. 21, no. 3, pp. 2831-2848, 2019.
[http://dx.doi.org/10.1109/COMST.2019.2907650]

[19]   T. Alladi, V. Chamola, J.J.P.C. Rodrigues, and S.A. Kozlov, "Blockchain in Smart Grids: A Review

on Different Use Cases", *Sensors (Basel),* vol. 19, no. 22, pp. 4862-4862, 2019.
[http://dx.doi.org/10.3390/s19224862] [PMID: 31717262]

[20]   G. Dileep, "A survey on smart grid technologies and applications", *Renew. Energy,* vol. 146, pp. 2589-2625, 2020.
[http://dx.doi.org/10.1016/j.renene.2019.08.092]

[21]   N.S. Nafi, K. Ahmed, M.A. Gregory, and M. Datta, "A survey of smart grid architectures, applications, benefits and standardization", *J. Netw. Comput. Appl.,* vol. 76, pp. 23-36, 2016.
[http://dx.doi.org/10.1016/j.jnca.2016.10.003]

[22]   N. Nidhi, D. Prasad, and V. Nath, "Different aspects of smart grid: An overview", *Nanoelectronics, Circuits and Communication Systems, Part of the Lecture Notes in Electrical Engineering,* vol. 511, Springer, pp. 451-456, 2019.

[23]   KE Mwangi, S Masupe, and J Mandu, "Modelling malware propagation on the internet of things using an agent-based approach on complex networks", *Jordanian Journal of Computers and Information Technology (JJCIT),* vol. 6, no. 1, pp. 26-40, 2020.

[24]   J.E. Sullivan, and D. Kamensky, "How cyber-attacks in Ukraine show the vulnerability of the U.S. power grid", *Electr. J.,* vol. 30, no. 3, pp. 30-35, 2017.
[http://dx.doi.org/10.1016/j.tej.2017.02.006]

[25]   HB Salameh, M Dhainat, and E Benkhelifa, "A Survey on Wireless Sensor Network-based IoT Designs for Gas Leakage Detection and Fire-fighting Applications", *Jordanian Journal of Computers and Information Technology (JJCIT),* vol. 5, no. 2, pp. 60-72, 2019.
[http://dx.doi.org/10.5455/jjcit.71-1550235278]

[26]   E. Manavalan, and K. Jayakrishna, "A review of Internet of Things (IoT) embedded sustainable supply chain for industry 4.0 requirements", *Comput. Ind. Eng.,* vol. 127, pp. 925-953, 2019.
[http://dx.doi.org/10.1016/j.cie.2018.11.030]

[27]   P.A. Pegoraro, A. Meloni, L. Atzori, P. Castello, and S. Sulis, "PMU-Based Distribution System State Estimation with Adaptive Accuracy Exploiting Local Decision Metrics and IoT Paradigm", *IEEE Trans. Instrum. Meas.,* vol. 66, no. 4, pp. 704-714, 2017.
[http://dx.doi.org/10.1109/TIM.2017.2657938]

[28]   A. Meloni, P.A. Pegoraro, L. Atzori, and S. Sulis, "An IoT architecture for wide-area measurement systems: A virtu- alised pmu-based approach", *Proc. of the IEEE In- ternational Energy Conference (ENERGYCON),* pp. 1-6, 2016.

[29]   I. Almomani, and K. Sundus, "The impact of mobility models on the performance of authentication services in wireless sensor networks", *Jordanian Journal of Computers and Information Technology,* vol. 06, no. 0, p. 1, 2020.
[http://dx.doi.org/10.5455/jjcit.71-1563658722]

[30]   NA Bakar, WMW Ramli, and NH Hassan, "The internet of things in healthcare: an overview, challenges and model plan for security risks management process", *Indonesian Journal of Electrical Engineering and Computer Science (IJEECS),* vol. 15, no. 1, pp. 414-420, 2019.

[31]   N. Zidková, M. Maryska, P. Doucek, and L. Nedomova, "Security of Wi-Fi As a Key Factor for IoT", *Hradec Economic Days,* 2020.

[32]   FI Salih, NAA Bakar, NH Hassan, F Yahya, N Kama, and J Shah, "IoT Security Risk Management Model for Healthcare Industry", *Malaysian J of Comp Science,* no. 3-9, pp. 131-144, 2019.
[http://dx.doi.org/10.22452/mjcs.sp2019no3.9]

[33]   R. Mahmoud, T. Yousuf, F. Aloul, and I. Zualkernan, "Internet of Things (IoT) Security: Current Status, Challenges and Prospective Measures", *Proc. of the 10th IEEE International Conference for Internet Technology and Secured Transactions (ICITST),* pp. 336-341, 2015.

[34]   J. Pacheco, and S. Hariri, "IoT Security Framework for Smart Cyber Infrastructures", *Proc. of the 1st IEEE International Workshops on Foundations and Applications of Self* Systems (FAS*W),* pp. 242-

247, 2016.
[http://dx.doi.org/10.1109/FAS-W.2016.58]

[35]   T. Alladi, V. Chamola, B. Sikdar, and K.K.R. Choo, "Consumer IoT: Security Vulnerability Case Studies and Solutions", *IEEE Consum. Electron. Mag.,* vol. 9, no. 2, pp. 17-25, 2020.
[http://dx.doi.org/10.1109/MCE.2019.2953740]

[36]   R. Kumar, R. Gupta, and S. Kumar, "IoT Security on Smart Grid: Threats and Mitigation Frameworks", *ECS Trans.,* vol. 107, no. 1, pp. 14623-14630, 2022.
[http://dx.doi.org/10.1149/10701.14623ecst]

[37]   R. Kumar, R. Gupta, S. Kumar, and N. Gupta, "IOT Security Framework Optimized Evaluation for Smart Grid", *International Journal of Electrical and Electronics Research,* vol. 12, no. 2, pp. 383-392, 2024.
[http://dx.doi.org/10.37391/ijeer.120208]

[38]   P. Srikantha, and D. Kundur, "A DER Attack-Mitigation Differential Game for Smart Grid Security Analysis", *IEEE Trans. Smart Grid,* vol. 7, no. 3, pp. 1476-1485, 2016.
[http://dx.doi.org/10.1109/TSG.2015.2466611]

CHAPTER 3

Towards the Assessment of Federations of Clouds

Bharat Chhabra[1,*] and **Shilpa Gupta**[2]

[1] *Department of Computer Science, Govt. College for Women, Karnal, India*

[2] *Department of Electronics and Communication Engineering, Chandigarh University, Mohali, India*

Abstract: The true driving force for the creation of a federation of clouds is a few fundamental characteristics, including the variety of infrastructures, interfaces, and different aims. The federation must ensure the application of certain standards and interfaces that allow secure and effective communication across heterogeneous entities in order to ensure seamless and helpful interaction between diverse components or entities of the various cloud providers. The federation has many commercial, legal, and technical aspects to focus on. Major features like resource provisioning, security, monitoring, *etc.* are suggested differently in various types of federations. This chapter analyzes a number of federation architectures on various important parameters with a view to highlighting their effect on participating cloud providers. Aspects related to Service Level Agreement management, QoS, Security, and Scheduling are also discussed in the same comparison framework.

Keywords: Federation of clouds, Federation architecture, FCM, ICAF, Layered architecture, SCF.

INTRODUCTION

Distributed computing has its latest descendant known as cloud computing. It is further advanced to Inter-Clouds, which is relatively more scattered geographically and hence more complex too.

One such definition has been given in a study [1] and it states that Inter-Cloud computing is "A cloud model that, for the purpose of guaranteeing service quality, such as the performance and availability of each service, allows on-demand reassignment of resources and transfer of workload through an interworking of cloud systems of different cloud providers based on the coordination of each consumer requirements for service quality with each providers SLA and use of standard interfaces".

* **Corresponding author Bharat Chhabra:** Department of Computer Science, Govt. College for Women, Karnal, India; E-mail: bharat.pnp@gmail.com

Neha Kishore, Pankaj Nanglia, Shilpa Gupta & Ashutosh Kumar Dubey (Eds.)

GICTF (Global Inter-Cloud Technology) Forum [2] has been working and issuing many use cases and functional requirements for Inter-Cloud Computing since long. Another important work has been published by DMTF (Distributed Management Task force) Inc [3, 4]. that has been publishing various white papers on the interoperability of clouds.

Classifying Inter-Cloud on the Basis of Participation of Cloud Providers

The above definition is a generalization of the environment and it is not conveying any hint about the originator of the venture i.e. Inter-cloud. There are many perceptions to understand the participation of different cloud providers in an inter-cloud. Such classification may also be given as:

Federation of Clouds

It is the scenario where multiple cloud providers join hands together on their will to form a bigger pool of services and infrastructure to share resources [5 - 7].

Multi-Cloud

It is the environment where multiple cloud providers work independently or used by some service or clients directly [6]. It is also recognized as "Sky-computing" and "hybrid-cloud" by different researchers.

There are a few important parameters that can make us easily understand that whether a particular cloud provider will become a part of federation or multi-cloud or none of these. Such factors are listed here as:

- Ownership
- Scale of operation
- Competitiveness

The answers to the above factors will definitely lead to the choice of going into federation or multi-cloud or skipping it all together. For example, if the ownership of a cloud is private then it may not choose to be a part volunteer federation to let its customers move on its competitor's platform. So, such cloud providers having private ownership may be least likely to be a part of federation of clouds. Moreover, being a part of federation will force the cloud provider to follow standard APIs so that workload may be easily migrated from/to own resources to other provider's infrastructure in the federation of course. Whereas, government owned clouds may participate in a federation of cloud to improve the scalability and QoS to public.

Understanding the Role of Inter-Cloud-Broker

The major distinction lies in the willingness to participate as it is present in the former scenario and absent in latter whereas falls under the same category i.e. Inter-Cloud. To properly exploit the features of having multiple clouds (connecting together in any manner) the role of inter-cloud-broker that acts on behalf of client becomes very vital. Its main job is to provision the resources from the bigger pool of resources and deploy the incoming tasks or applications. Few crucial jobs besides many others that are expected to be performed by inter-cloud-broker are:

- Provisioning of resources across the clouds.
- Allocation and de-allocation of resources
- Scheduling
- Load-balancing
- Orthogonal resource-sharing

ARCHITECTURES IN INTER-CLOUDS

The architectures introduced in the previous section namely Multi Clouds and Federation of Clouds are very different in their approach. This leads to their varied operational needs leading to diverse challenges at the broker level for due implementation of these architectures. To understand these challenges, it is firstly important to understand the architectural intricacies of both the models as described in the next section.

Architecture of "Federation of Clouds"

The architecture of this type of organization in Inter-Clouds as shown in Fig. (1) and Deployment models (Fig. 2) may be represented as [8].

Centralized

In such an architecture of federation, a centralized entity called an Inter-Cloud-Broker (ICB) performs all the functions related to resource provisioning, allocation and workload distribution. All cloud providers that participate in the federation are to register the information of resources in a central repository with an inter-cloud-broker.

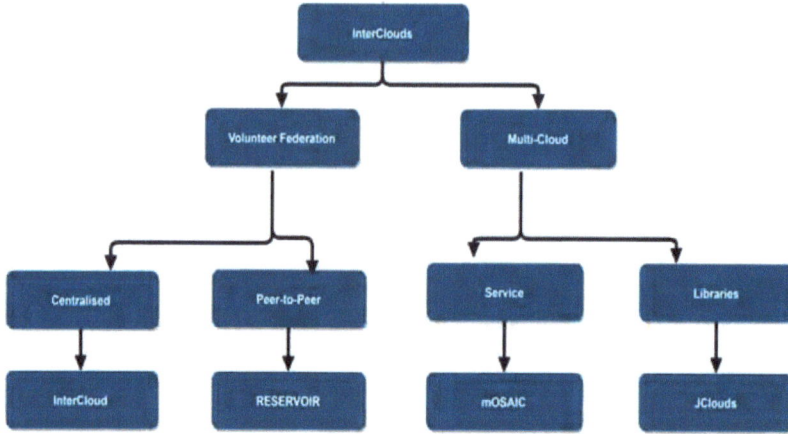

Fig. (1). Classification of architectures of federation of clouds.

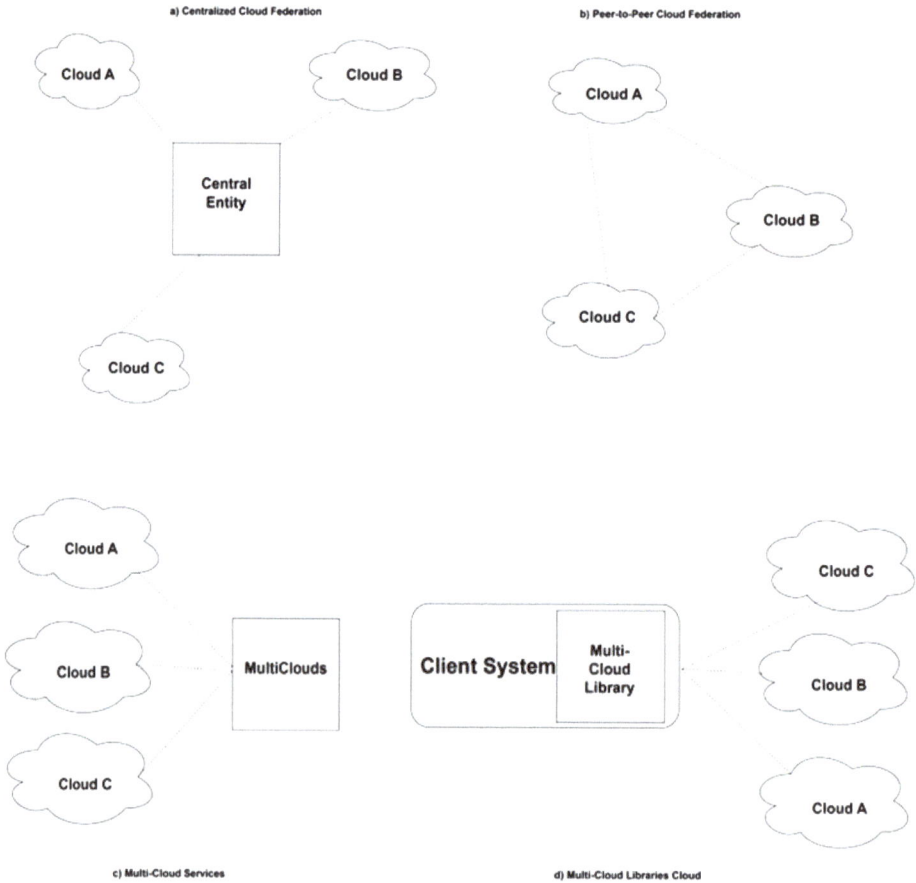

Fig. (2). Deployments of inter-clouds.

Peer-to-Peer

In such architecture of federation, the cloud provider's local cloud broker module interacts directly with other participants' local broker for the mediation of services and resources.

The Architecture of "Multi-Clouds"

The architecture of this type of organization in Inter-clouds shown in Fig. (**2**) may be classified as:

Collection of Services

In such architecture, the services being offered follow a set of rules detailed in the Service Level Agreement (SLA) in the form of if-then rules. These rules of SLA govern the provisioning mechanism, allocation and deployment schemes in the background.

Collection of Libraries

In such architecture of Multi-Clouds, a set of libraries is made available to facilitate the usage of resources from multiple cloud providers in a symmetric manner. The task of provisioning and allocation is left to the application brokers itself.

BROKERING MECHANISM IN INTER-CLOUD

It is quite evident from the previous section that the brokering mechanism is very different in both the architectures of Inter-cloud. In fact, the entity doing the brokering plays a crucial role in the success of the inter-cloud and so its placement in both architectures defines the way it shall behave. In the Federation of Clouds, this task is taken care of by a centralized broker on behalf of all participating cloud providers whereas, in case of multi-cloud services, the brokering is managed by the service that is providing access to the cloud. In fact, an external entity (in multi-clouds) managing the brokering is more transparent to the application. Such externally managed brokering mechanisms are represented in Fig. (**3**) and those are divided into the following types.

Service-Level-Agreement

The brokering module is specified in terms of certain rules in SLA along with its constraints and goals. Each cloud provider participating in multi-cloud has to compulsorily honor the terms of SLA. The uniqueness of this mechanism lies in the fact that it is fully transparent since all the management is done in the background and not visible to application developers. The development part becomes lighter and easier as this brokering is managed by the service provider behind the curtain.

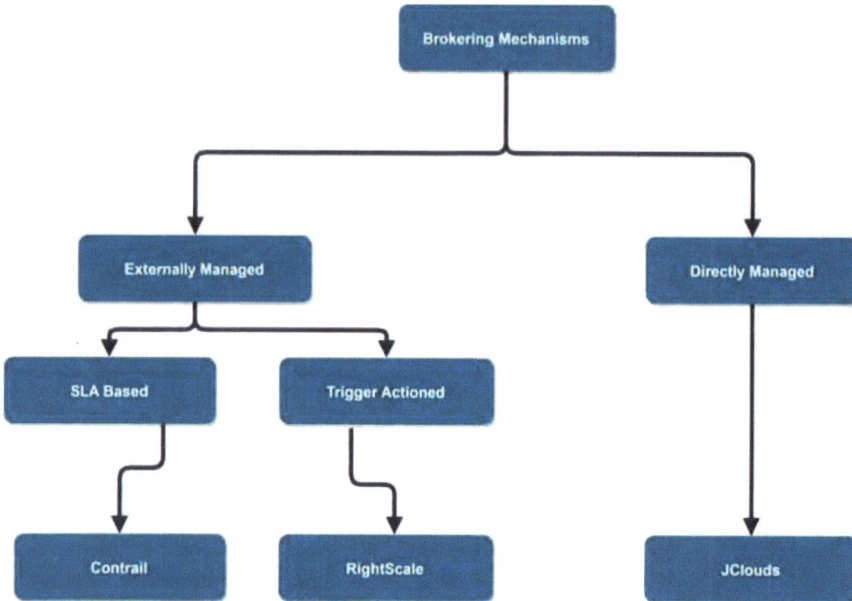

Fig. (3). Brokering mechanisms.

However, it also needs a good level of trust between the client and provider as the client does not have any control over the provisioning of resources across the cloud. Generally, a non-negotiable SLA is offered to clients having a standard format where the client does not have much to alter according to the requirements of tasks. The Clouds Standards Customer Council also stated that the SLA being offered nowadays is immature and there is a need for mature SLAs to enable the customer to specify its provisioning requirements and other needs [9].

Trigger-Action

In this mechanism, the brokering logic is detailed in terms of cause-effect relationship rules. Certain events/actions are triggered based on certain conditions i.e. an action is initiated at the occurrence of a certain condition. For example, if the number of connections to a VM goes beyond a certain limit say 10, then the amount of memory allocated to the VM may be doubled or a new VM may be instantiated immediately in order to maintain the performance. This trigger-action approach is less transparent, hence offers more control to its customer and enables it to specify its own provisioning policies and scalability rules.

A federation with heterogeneous clouds consists of multiple cloud providers having different hardware/software configurations each having multiple Data Centers. These data centers must interact and communicate with one another regularly and effectively. As a result, the complexity that needs to be controlled is increased. Federated clouds do improve service availability and fault resilience, but they also present some difficulties with load balancing, autonomic scaling, SLA management, and other issues.

To maximize its gains, a cloud provider needs to carefully assess and examine the internal functioning and details of the federation before joining it. This requires an analysis of the existing federations from various perspectives. Most significant of these are interoperability [10, 11], scalability, the type of architecture, workload balancing, security, privacy, and monitoring.

RELATED WORK

Conceptual model of federation was developed through the mapping of expected services of a federation into components. The model has been presented as a layered architecture. This model grouped together logical components into segments, which are then assembled into layers. The conceptual model architecture contained 4 layers namely business, logical, repository and communication. The first layer comprised segments pertaining to service contracts. The logical layer deals with the technical details such as workload balancing, privacy and monitoring. The repository layer contains the virtual and physical resources of federation. This layered architecture is shown in Fig. (**4**).

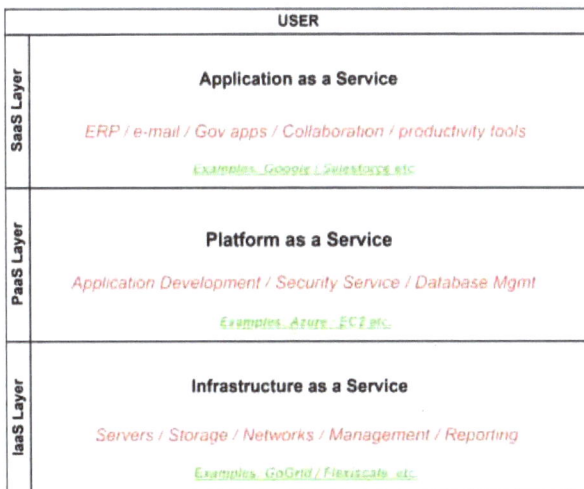

Fig. (4). Layered architecture of cloud computing.

The layered architecture of federation is a layered Cloud service model of software (SaaS), platform (PaaS), and infrastructure (IaaS) that enables multiple independent Clouds to form a federation. The layered architecture allows a cloud provider to participate in inter-cloud federation at any of the service layers. In this Cloud federation, negotiation between two providers at the same layer takes place with the help of a few well-defined sets of parameters. The benefit of running an application that can be supported by several providers implementing different sections of the functionality is really provided by the layered architecture of federation.

The authors explained the process of establishing the federation between participating cloud providers and the process of delegation of tasks to lower layers of the same provider.

The Simple Cloud Federation (SCF) architecture is the hybrid cloud architecture. Considering that it took advantage of the hierarchical association features already available in the majority of well-known current cloud systems, it differs from previous federation models. The authors developed the SCF architecture to enable federation between an OpenStack cloud and Amazon EC2.

FCM architecture of federation has been proposed [1] that offered a single entry point to cloud federation through Generic Metabroker. This meta-brokering component tenders the interface to cloud providers to connect to the federation. In order to choose an appropriate cloud broker for job submission and execution, the GMBS consults the generic repository. The job-resource mapping is done using both static information from the FCM Repository and dynamic data from particular deployment metrics that the cloud brokers have collected. For the service requests received from the federation, cloud providers can manage the used virtual machines with their cloud brokers.

ICAF of the Federation of Clouds presented the research work to develop the Intercloud Architecture Framework (ICAF) that proposed to mitigate the problems in multi-provider multi-domain heterogeneous cloud-based infrastructure services and applications' integration and interoperability. The architecture supports the communication between cloud applications at different layers and different sites. It also supports and controls the infrastructure services to achieve run-time optimization. It has been presented as a layered service model architecture with seven horizontal layers and three vertical layers to facilitate all intra and inter-cloud service provisioning.

CRITERIA FOR COMPARISON OF ARCHITECTURES

For the appropriate working of the federation, its framework should be robust. Various cloud federation architectures have been proposed so far. This paper compares these architectures on a number of vital parameters.

Architecture

The architecture of the federation of clouds has many possible variants including CCOA: Cloud Computing Open Architecture [12], InterCloud [8, 13], and a few others. These available architectures may be compared on a number of parameters that are important for drawing a thin line between them. A few major types of architectures are:

Conceptual Model Architecture

Conceptual model architecture for the federation of clouds is a layered architecture containing four different layers. Each layer handles various functions that are logically grouped. This architecture has no centralized governance system for the federation. Moreover, this architecture assumes the independence of underlying technology for virtualization. The special feature of this architecture is that it supports interoperability with help of some external interfaces and operational interfaces [11].

Layered Architecture

As shown in Fig. (**5**), it is composed of three layers, having no centralized governance system to manage the federation.

The most distinguishing feature of this federation is that the brokering decisions are taken at each layer independently. Hence the objectives at each layer are different at each layer and the focus is on improving the performance at each layer. Interoperability issues are quite complex in this architecture since federation is possible at each layer.

SCF Architecture

SCF architecture is a hierarchical architecture for hybrid clouds in particular. There is no centralized broker in this architecture as represented in Fig. (**6**). To regulate the transfer onto the foreign cloud, brokering regulations may be established in the main cloud. The association with another cloud is established to form a federation on an ad-hoc basis. It is up to the primary cloud to decide when to establish this federation.

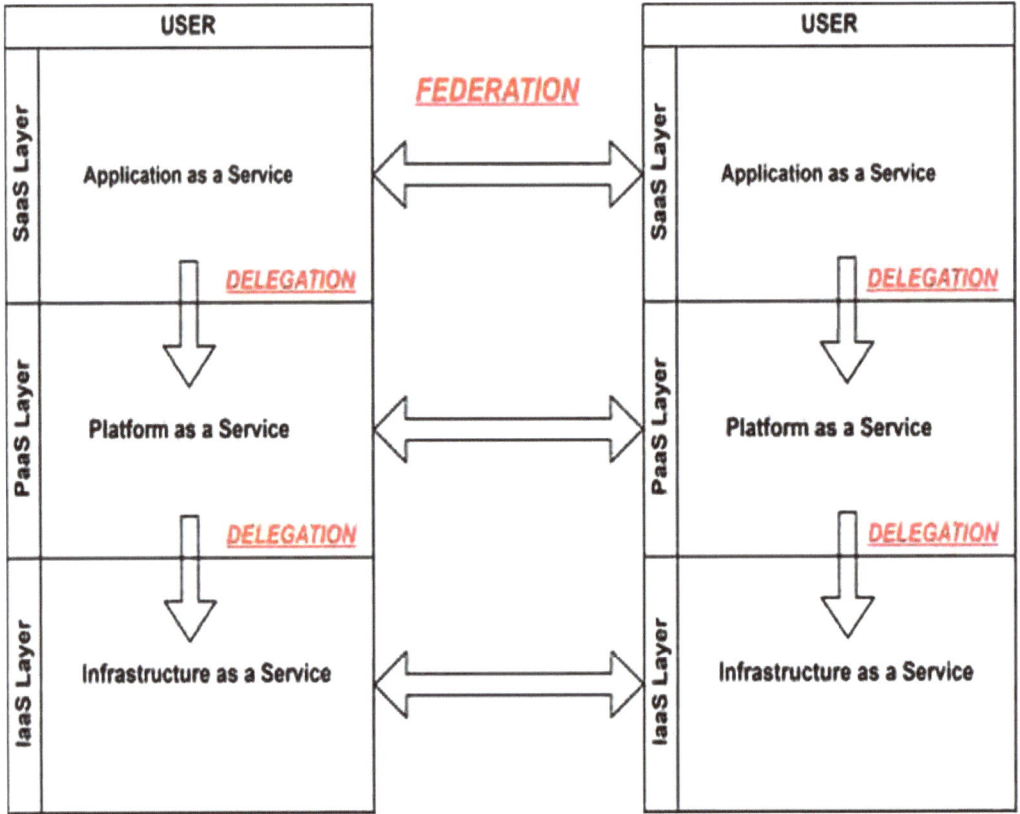

Fig. (5). Three-layered architecture of the federation of clouds.

The FCM Architecture

The FCM architecture is a centralized federation of clouds with a key element known as GMBS [1]. As represented in Fig. (7), this GMBS is a single point of entry that serves as the highest-level broker to choose the computing environment to which the task request will be forwarded. This design does not mandate that every CSP embraces interoperability standards; instead, a Global-Meta-Brokr-System handles this. By using a transparent technique, inter-cloud communication can be made possible without the use of additional software. The management of the federation is the duty of this central GMBS component of FCM. Both the global and local levels of the architecture are managed. At both levels, interoperability is also taken into consideration.

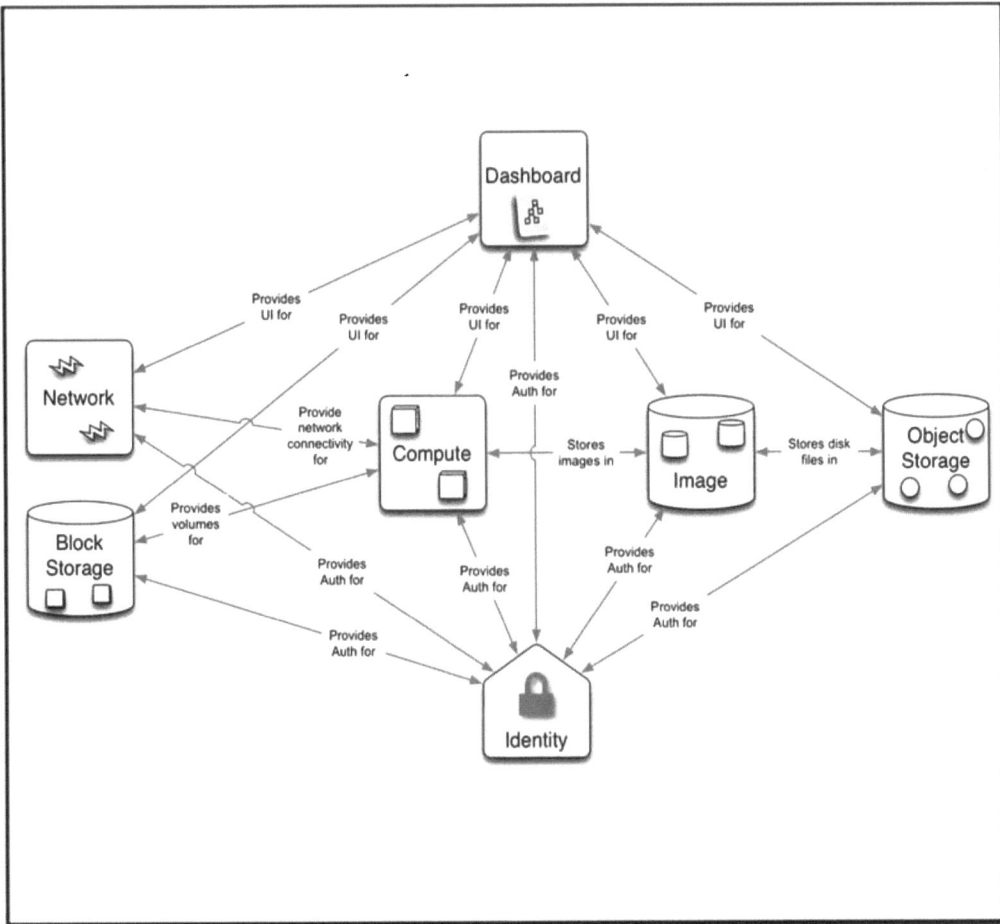

Fig. (6). SCF Architecture of Federation of Clouds.

Authors in a study [14] have amended FCM architecture with the addition of a few rule-based features that enabled the architecture to be self-adaptable. This rule-based repository is known as its Knowledge Management (KM) solution.

• The Cloud Service Model (CSM), InterCloud Control and Management Plane (ICCMP), InterCloud Federation Framework (ICFF) as shown in Fig. (8), and InterCloud Operation Framework (ICOF) are the four main parts that make up the ICAF architecture.

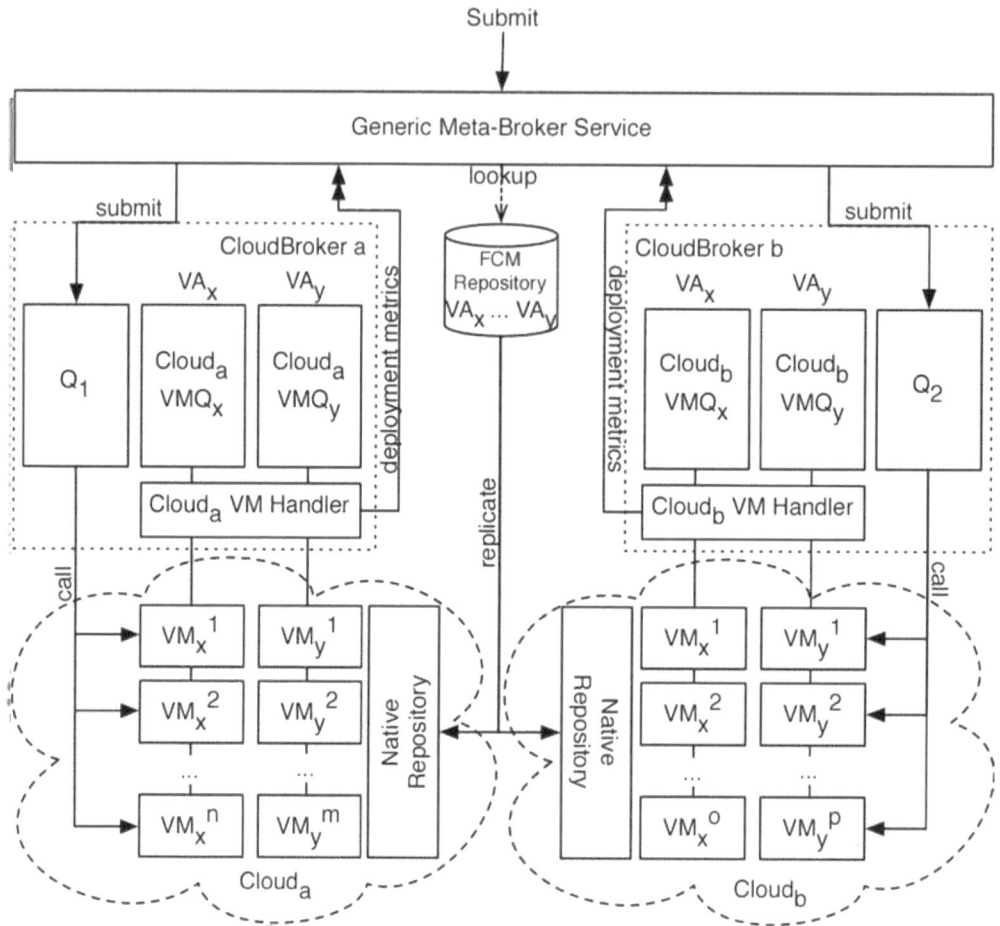

Fig. (7). FCM Architecture.

Each layer, plane, or component provides a unique dedicated mechanism for carrying out specific activities. CSM is used for integrating and interacting with cloud services. Inter-application signaling, synchronization, session management, configuration, and monitoring are all part of the inter-cloud control and management protocol (ICCMP). Assuming sufficient gateway or federation services, the ICFF ought to facilitate federation at the level of services, business applications, semantics, and namespaces. ICOF features business process, SLA management, and accounting to allow the use of multi-provider infrastructure.

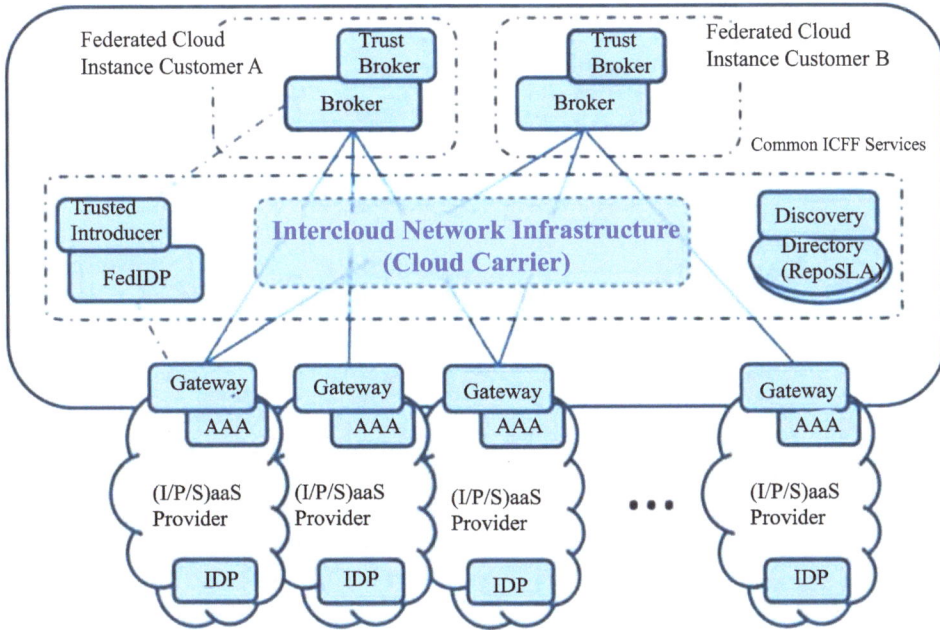

Fig. (8). ICAF-ICFF Architecture.

SLA and QoS Monitoring

QoS stands for Quality of Service [15], and in the context of the federation of clouds, it refers to the ability to deliver specific performance and service guarantees to users and applications. QoS is essential for ensuring that the required levels of performance, reliability, and availability are maintained within the federated cloud environment. QoS comprises the following factors in the federation of clouds:

- Performance: QoS aims to provide consistent and predictable performance in terms of response time, throughput, and latency for applications and services. It involves optimizing resource allocation, load balancing, and scheduling mechanisms to meet the performance requirements defined by service-level agreements (SLAs).
- Reliability: QoS ensures that the federated cloud environment is reliable and resilient. It includes measures such as fault tolerance, data replication, backup and recovery mechanisms, and high availability configurations to minimize service disruptions and ensure continuous service delivery [15].
- Availability: QoS guarantees the availability of resources and services within the federation. It involves redundancy, failover mechanisms, and load balancing

strategies to distribute workloads and avoid single points of failure. The goal is to provide uninterrupted access to services, even in the event of failures or disruptions [16].

- Scalability: QoS ensures that the federated cloud environment can scale resources and services based on demand. It involves dynamic resource provisioning, elasticity, and auto-scaling capabilities to handle varying workloads and accommodate increased resource requirements as needed.
- Security: QoS includes security measures and controls to protect data, applications, and resources within the federation. This involves implementing robust authentication mechanisms [4], access controls, encryption, and compliance with relevant security standards and regulations. Maintaining a high level of security is crucial for preserving the integrity and confidentiality of user data.
- Service-Level Agreements (SLAs): QoS in the federation of clouds is often defined through SLAs. SLAs establish the expected levels of performance, availability, and reliability that the cloud providers commit to delivering. QoS monitoring and reporting mechanisms help ensure compliance with SLAs and enable transparency and accountability in service delivery.

By providing consistent performance, reliability, availability, scalability, and security, QoS in the federation of clouds aims to meet the needs and expectations of users and applications. It establishes a foundation of trust and ensures that the federated cloud environment can effectively support a wide range of services and workloads.

- Business layer of conceptual architecture deals with QoS, SLA and federated contract establishment that may contain provisions for penalties. The monitoring segment of this architecture offers monitoring related to the federation itself called global monitoring. Monitoring as a service can also be offered to customers. Security of the monitoring system itself is also regarded as important and of high significance.

- The Cloud Federation's Layered Architecture performs monitoring of the whole federation by tracking data about its virtual as well as physical resources. A few key metrics compute execution time and transmission costs and then share this information with other clouds that are being courted for federation. This federation's monitoring component preserves the log and metric data, and this data is utilized to govern the federation.

- SCF architecture does not introduce any new component in its architecture for monitoring rather it is up to the primary cloud to implement or do the monitoring of the service or resource to the extent possible in its domain.

• The FCM architecture continuously monitors and evaluates the data center's infrastructure, services' behavior, security, utilization of energy, SLA implementation, and resource performance. At a more fundamental level, a separate tracking component is used to keep track of each VM's memory, disc, network, CPU load, typical VM deployment time, and other running resources' present conditions.

• With the help of SALMon, a unified surveillance strategy, FCM has been improved as stated in [14]. The monitoring mechanism keeps track of the QoS and dependability of every service in existence. Each service is quality-checked by the SALMon framework, which then provides this information to the GMBS IS agent. SALMon is designed to be sufficiently compact to run on virtual machines and be installed in each cloud that FCM manages. The reliability of a specific cloud is expressed by SALMon using the reaction times of a few fundamental methods including ping, stress CPU, stressinoutbandwidth, and stressoutputbandwidth. The monitored values are updated by the GMBS IS agent.

• The ICAF addresses the need to monitor various services and physical as well as virtual resources with the help of its dedicated component ICCMP. ICCMP is responsible for run-time optimization of physical and virtual infrastructure. To accomplish this, various metrics are maintained for each of the resource types, and compared with benchmarks at appropriate times. A slight deviation in performance levels of any resource type triggers the fine-tuning of that resource type.

• To ensure the consistent performance of the federation, the monitoring subsystem of ICAF records the historical events and performance metrics that can be reported for automatic configuration management of clouds. The Cloud Service Registry process is monitored for authentication. SLA enforcement is monitored and compared with acclaimed performance for control purpose.

• The Cloud Configuration Management subsystem of the Cloud Management Plane is responsible for retrieving the cloud configuration data and offering a method for communicating feedback to the management system. Information on network status and monitoring is also included in this feedback. NETCONF, CLI, and SNMP are the protocols that are employed for this.

Scheduling and Load Balancing

Scheduling and load balancing are two important aspects in the context of the federation of clouds. They involve the efficient allocation of resources and the distribution of workloads across multiple cloud service providers (CSPs) within the federation.

Load balancing in the federation of clouds involves the distribution of incoming requests across multiple CSPs to ensure optimal resource utilization, avoid resource bottlenecks, and improve overall system performance. Load balancing algorithms aim to evenly distribute the workload while considering factors such as resource availability, network latency, and current load conditions.

• Repository layer of the conceptual architecture deals with the resources and resource segment of this architecture making available all types of physical and virtual assets. It also performs accounting of resources' offers and consumption. The orchestration segment deals with workload balancing. The workload balancing decision is based on factors like data locality, topology, and architecture of the cloud/federation.

• The layered architecture of cloud federation has empowered each layer (SaaS/PaaS/IaaS) to establish the federation that undertakes the scheduling of jobs at the IaaS layer and the process of workload consolidation is also proposed in the layered architecture at the IaaS layer only that too with an objective of energy efficiency.

• Since SCF architecture does not introduce any broker service and relies on individual capabilities of participating clouds to approach a foreign cloud as it suits its individual needs and does not propose to implement any global policy to maintain a balanced workload on all clouds of the federation.

• The Generalised Meta Broker Service in the FCM architecture is in charge of scheduling all requests. Using the Basic Property Description Language (BPDL), all scheduling-related properties, including the anticipated availability time for each VM, are kept up to date across all cloud brokers in the "PerformanceMetrics" field. The information thereby gathered is utilised to rate all cloud brokers, and based on this rank, the best scheduling alternatives are selected for each input work, the best cloud broker is chosen using a consistently tracked sequence of statistics gathering and rank calculation.

• This GMBS is supported by an integrated Knowledge Management system that takes care of appropriate load distribution amongst participating clouds to achieve objectives like better response time, throughput, *etc.* For the purpose of identifying instances when the architecture exhibits poor behaviour, this KM subsystem aggregates the monitoring metrics, such as throughput, awt, cvm ratio, load, and cost. When this happens, the KM subsystem may advise self-sufficient actions like expanding or contracting or moving the VM queue or VA storage, or it may even prompt GMBS-level service call rescheduling.

• In ICAF, the scheduling and load-balancing process is controlled on the basis of a few established metrics. During the execution phase, monitoring is performed in real-time for optimized resource usage and cloud performance. The performance of VMs is monitored by calculating the established metrics. These metrics suggest the required level of virtual resource orchestration that is done through VM migration or load balancing. This consistent monitoring of VMs provides extensive support for runtime resource optimization i.e. to assess the need for additional VM's introduction, alteration, or deletion.

Security and Privacy

Security in the federation of clouds refers to the measures and practices employed to protect the confidentiality, integrity, and availability of data and resources in a federated cloud computing environment [17, 18]. In a federated cloud, multiple independent cloud service providers (CSPs) collaborate to provide a unified cloud infrastructure and services to users.

Here are some key aspects of security in the federation of clouds:

a) Data Confidentiality: This ensures that data is protected from unauthorized access or disclosure during storage, processing, and transmission [20]. Encryption techniques, access controls, and secure communication protocols are employed to maintain data confidentiality.

b) Data Integrity: This ensures that data remains unchanged and uncorrupted throughout its lifecycle [19]. Techniques such as data validation, checksums, digital signatures, and secure storage mechanisms are used to maintain data integrity.

c) Authentication and Access Control: This verifies the identity of users and entities accessing the federated cloud and enforcing appropriate access controls [17]. This includes authentication mechanisms like passwords, multi-factor authentication, and integration with identity and access management systems.

d) Network Security: It protects the network infrastructure of the federated cloud against unauthorized access, attacks, and data breaches [18]. Firewalls, intrusion detection and prevention systems, virtual private networks (VPNs), and secure network protocols are employed to safeguard the network.

e) Secure Virtualization: Ensuring the security of virtualized resources and hypervisors used in the federated cloud environment [19]. Isolation between different virtual machines (VMs), secure hypervisor configurations, and vulner-

-ability management practices are employed to mitigate virtualization-related security risks.

f) Incident Response and Monitoring: Implementing systems and processes to detect and respond to security incidents promptly. Security monitoring tools, log analysis, real-time alerting, and incident response plans are important components of a robust security framework [21].

g) Compliance and Auditing: This ensures adherence to regulatory and industry-specific security requirements and conducting regular security audits to ensure compliance [22]. This includes data privacy regulations, such as GDPR, HIPAA, or PCI-DSS, depending on the nature of the data and the industry.

h) Secure Interoperability: This ensures secure communication and data exchange between different cloud providers within the federation [10, 11]. Standardized protocols, secure APIs, and secure data transfer mechanisms are used to facilitate interoperability while maintaining security.

i) Trust and Legal Considerations: This establishes trust among the cloud providers in the federation, defines legal agreements, and addresses liability, data ownership, and contractual aspects [22]. Clear policies and agreements help build trust and define the responsibilities of each party.

Overall, security in the federation of clouds requires a comprehensive approach that encompasses various layers of the cloud infrastructure, including data, networks, virtualization, and user access, while adhering to relevant regulations and industry best practices.

The conceptual architecture of cloud federation includes a security segment that is dedicated to implementing security policies, and authentication schemes, avoiding internal/external attacks, and mapping the borders. Another segment called the communication segment of this architecture encloses the components and protocols that deal with only inter-cloud communication but not intra-cloud communication.

Security in the context of execution environment in layered architecture is taken care of at thePaaS layer due to the existence of platform differences among participating cloud providers.

SCF architecture does not define any dedicated segment or component to take care of privacy and security management of the federation rather it relies on each individual cloud provider to enforce their own internal or proprietary security mechanisms to ensure the connection with any foreign cloud.

The cloud management plane of ICAF offers a separate dedicated mechanism for service registration and discovery that helps to maintain proper security and privacy of the participating clouds in the federation besides making it transparent to the consumer. Monitoring in ICAF is also extended to the inter-cloud connection requirements. It provides the mechanism for monitoring delay-jitte--loss and UDP/TCP flow across the clouds, ensuring a secure transport channel.

CONCLUSION AND FUTURE WORK

In this paper, a few pioneer reference architectures of the federation of clouds have been investigated that offer autonomous cloud management and also consist of various specialized solutions for providing mandatory services like security, monitoring, QoS, load balancing, *etc.* It has been observed that all architectures have their unique way of providing services and managing the federations. The focus of the conceptual model of the federation is to logically categorize all the required functionalities into a few segments and to make these segments work in a coordinated fashion. The layered architecture of federation does welcome small cloud providers in the federation that provide services at one of the layers only but it does not attend to the privacy or security issues in federation architecture. SCF architecture of federation has a hierarchical style of establishing the federation with another cloud on an ad-hoc basis with a primary cloud to tackle security, monitoring, load balancing, *etc.* FCM architecture of federation has a centrally GMBS-dependent structure that controls all of the major services. It has also been enhanced with the help of the KM system to attain autonomous management of the federation. Another enhancement of FCM isthat it has incorporated the SLAM on system to enable effective monitoring of the federation. ICAF has been proposed to efficiently deliver all expected services through one of its core components. Some of the ICAF's services, including tracking, are highly flexible and start with service registry and discovery, continue with SLA realization, and continue while execution is taking place at the moment to track cloud performance. In order to improve performance, such real-time monitoring may lead to the migration of virtual machines or tweaking of other run-time resources.

The investigation in this chapter paves the path for any cloud provider to pinpoint the delivery of essential features of existing federations of clouds. There will undoubtedly be more work done to close the gaps that are identified throughout this study, including those in task distribution, security, and privacy.

REFERENCES

[1]　A. Marosi, G. Kecskemeti, A. Kertesz, and P. Kacsuk, "FCM: An architecture for integrating IaaS cloud systems", *Proceedings of the Second International Conference on Cloud Computing, GRIDs, and Virtualization* Rome, Italy, pp. 43, 7-12, 2011.

[2]　GICTF (Global Inter-Cloud Technology) Forum, "Use Cases and Functional Requirements for Inter-

Cloud Computing", August 2010.

[3]   DMTF (Distributed Management Task force) Inc., White Paper "Interoperable Clouds",2009.

[4]   GICTF (Global Inter-Cloud Technology) Forum, White Paper "Technical Requirements for Supporting the Intercloud Networking", 2012.

[5]   B. Rochwerger, D. Breitgand, E. Levy, A. Galis, K. Nagin, I.M. Llorente, R. Montero, Y. Wolfsthal, E. Elmroth, J. Cáceres, M. Ben-Yehuda, W. Emmerich, and F. Galán, "The Reservoir model and architecture for open federated cloud computing", *IBM J. Res. Develop.*, vol. 53, no. 4, pp. 4:1-4:11, 2009.
[http://dx.doi.org/10.1147/JRD.2009.5429058]

[6]   A.J. Ferrer, F. Hernández, J. Tordsson, E. Elmroth, A. Ali-Eldin, C. Zsigri, R. Sirvent, J. Guitart, R.M. Badia, K. Djemame, W. Ziegler, T. Dimitrakos, S.K. Nair, G. Kousiouris, K. Konstanteli, T. Varvarigou, B. Hudzia, A. Kipp, S. Wesner, M. Corrales, N. Forgó, T. Sharif, and C. Sheridan, "OPTIMIS: A holistic approach to cloud service provisioning", *Future Gener. Comput. Syst.*, vol. 28, no. 1, pp. 66-77, 2012.
[http://dx.doi.org/10.1016/j.future.2011.05.022]

[7]   Available from: http://www.arjuna.com/what-is-federation [last accessed 14 June 2012].

[8]   Grozev Nikolay and BuyyaRajkumar, "Inter-Cloud Architectures and Application Brokering: Taxonomy and Survey", "Software – Practices and Experience – Published in Willey InterScience", 1-22, 2012.

[9]   CSCC Workgroup. Practical guide to cloud service level agreements version 1.0. Technical Report, Cloud Standards Customer Council (CSCC), 2012.

[10]   Tahereh Nodehi et. al, "On MDA-SOA based Intercloud Interoperability framework", Computational Methods in Social Sciences" Vol. I, Issue 1/2013, ISSN 2344-1232.

[11]   SNE (System and Network Engineering Group), "Intercloud Architecture for Interoperability and integration", Release 2, Draft Version 0.6, 2013.

[12]   L-J. Zhang, and Q. Zhou, "CCOA: Cloud Computing Open Architecture", *IEEE International Conference on Web Services,* pp. 607-616, 2009.

[13]   BuyyaRajkumar, Ranjan Rajiv and Calheiros N. Rodrigo, "InterCloud: Utility-Oriented Federation of Cloud Computing Environments for Scaling of Application Services", *10th International Conference on Algorithms and Architectures for Parallel Processing*, pp 13-31, 2010.

[14]   G. Kecskemeti, M. Maurer, I. Brandic, A. Kertesz, Z. Nemeth, and S. Dustdar, "Facilitating self-adaptable Inter-Cloud management", In: *Proceedings of the 20th Euromicro International Conference on Parallel, Distributed and NetworkBased Processing (PDP 2012)* IEEE: Munich, Germany, 2012, pp. 575-582.
[http://dx.doi.org/10.1109/PDP.2012.41]

[15]   N. Saini, S. Gupta, and B. Thakral, "Multistage interconnection networks a review", *Int. J. Electron. Commun. Technol,* vol. 4, no. 3, pp. 1-4, 2013.

[16]   S. Gupta, and G.L. Pahuja, "A review on gamma interconnection network", *International Journal of Computational Systems Engineering,* vol. 5, no. 3, pp. 137-151, 2019.
[http://dx.doi.org/10.1504/IJCSYSE.2019.10022446]

[17]   David Nunez et. al, "Identity Management Challenges for Intercloud Applications", STA 2011 workshop, CCIS 187, pp-198-204, Springer 2011.

[18]   W. Li, and L. Ping, "Trust model to enhance security and interoperability of cloud environment", *Cloud Computing,* pp. 69-79, 2009.
[http://dx.doi.org/10.1007/978-3-642-10665-1_7]

[19]   H. Tsuda, A. Matsuo, K. Abiru, and T. Hasebe, "Inter-Cloud Data Security for secure Cloud-based Business Collaborations", *FUJISTU Sci. Tech. J.,* vol. 48, no. 2, pp. 169-176, 2012.

[20]   MacDermott, Áine & Shi, Qi & Merabti, Madjid & Kifayat, Kashif. Security as a Service for a Cloud Federation. 2014.

[21]   M. Rak, M. Ficco, J. Luna, H. Ghani, N. Suri, S. Panica, and D. Petcu, "Security issues in cloud federations", In: *Achieving Federated and Self-Manageable Cloud Infrastructures: Theory and Practice*, 2012, pp. 176-194.
[http://dx.doi.org/10.4018/978-1-4666-1631-8.ch010]

[22]   K. Bernsmed, M.G. Jaatun, P.H. Meland, and A. Undheim, "Thunder in the clouds: Security challenges and solutions for federated clouds", *4ᵗʰ IEEE International Conference on Cloud Computing Technology and Science Proceedings,* pp. 113-120, 2012.
[http://dx.doi.org/10.1109/CloudCom.2012.6427547]

CHAPTER 4

Challenges in Digital Payments and Financial Cyber Frauds in Rural India

Rahul Rajput[1,*] and **Bindu Thakral**[1]

[1] *Sushant University, Gurugram, Haryana, India*

Abstract: Improving digital payment trends in rural India is crucial given the growing impact of ICT penetration, demonetization, and digital activities for small businesses in rural sectors. The shift to digital payments can offer benefits such as transaction transparency, reducing parallel economy, and improving ease of doing business. Although various digital wallets such as Paytm, Mobikwik, and PhonePe have been introduced and the government has launched UPI solutions like the BHIM app, rural banking consumers still struggle to embrace digital payments due to the lack of digital literacy. India has a large rural population, but only a small percentage is digitally literate, hindering digital payment adoption. This research study examines the significance of digital literacy in the current banking environment, focusing on issues, opportunities, and difficulties related to the adoption of digital payments in the rural banking sector.

Keywords: Digital transactions, Demonetization, Digital payments, Digital divide, Government, PoS, Rural, Rural india.

INTRODUCTION

Even in rural areas, the Indian government has been pushing for a cashless society and digital payments. The COVID-19 outbreak intensified this endeavor and prompted the Ministry of Electronics and IT (MeitY) to introduce the "Digital Finance for Rural India" scheme, which aims to increase awareness and access to digital financial services through Common Service Centers (CSCs). The endeavor to spread awareness of services like IMPS, UPI, and Bank PoS machines received significant funding from the government. Through projects like the Bharat Net Project, the Prime Minister's 2015-launched "Digital India" initiative sought to guarantee that all residents, especially those living in rural areas, have access to

---

* **Corresponding author Rahul Rajput:** Sushant University, Gurugram, Haryana, India;
E-mail: rahulrajput5840565@gmail.com

Neha Kishore, Pankaj Nanglia, Shilpa Gupta & Ashutosh Kumar Dubey (Eds.)

government services online. The PM Jan DhanYojana, DBT, Atal Pension Yojana, and other programs (such as the introduction of RuPay cards) are encouraging digital literacy and empowerment where more than 55% of users are female. According to a statement made by the Ministry of Finance on December 24, 2021, 44.12 crore accounts under PM Jan DhanYojana were open as of December 15, 2021.

A total of INR 171,873.45 crore has been deposited into the accounts of 46.05 billion beneficiaries who have been banked thus far. In addition, 6.55 lakh 'Bank Mitras' offer branchless banking services across the nation [1 - 3].

The old system, which includes payment options like cheques, withdrawals, drafts, money orders, letters of credit, and traveler's checks, is being replaced with the digital system. Current Payment systems are electronic payment systems that employ computers and the internet for a variety of purposes empowering the usage of digital wallets and UPI platforms. The traditional system has some shortcomings and inefficiencies that the digital payment systems can address, which is one of the main drivers for this transition. The 'Ministry of Electronics and Information Technology' defines digital literacy as the capability of individuals and communities to comprehend and apply digital technology for useful purposes in daily life. Anyone who can operate a computer, laptop, tablet, smartphone, and other IT resources is considered to have digital literacy. According to this definition, we define a household as being digitally literate if at least one member can operate a computer and have access to the internet (for those who are five years old and older). This criterion led us to the conclusion that just 38% of Indian households are digitally literate. Digital literacy in urban regions is higher than in rural areas, where it is only 25% compared to 61% for their urban counterparts.

According to some estimates, India has a meager 6.5% computer literacy rate. Therefore, it is essential to recognize the challenges preventing the development of digital payment systems in rural India as well as the abuse of rural Indians' bank accounts in online financial fraud [4 - 6].

DIGITAL PAYMENT METHODS

The government's Digital India programme aims to improve India's digital infrastructure and create a wealthier, knowledge-based society. The goal of the Digital India programme is to use technology to advance business practices in India. This covers the application of anonymous, paperless, and cashless techniques. In India, there are several options to pay with digital devices that decrease the need for paying with cash-making in India a society that uses less cash.

BANKING CARDS (DEBIT/CREDIT/CASH/TRAVEL/OTHERS)

Banking cards, including debit, credit and pre-paid cards, provide more safety, convenience, and control to consumers *vis-à-vis* any other payment mode. These cards enhance safety by offering two-factor authentication, such as a secure PIN and an OTP, for safer transactions. Card payment systems include RuPay, Visa, and MasterCard, among others. People can make purchases with payment cards using multiple modes such as in-person, over the phone, online, through mail-order catalogues, and at retail establishments. These enable transactions while saving both customer's and merchant's time and money [7].

UNSTRUCTURED SUPPLEMENTARY SERVICE DATA (USSD)

Payment Service *99# is one such inventive solution that uses the Unstructured Supplementary Service Data (USSD). Without the need for a mobile internet data facility, this service makes it possible to conduct mobile banking transactions on simple feature mobile phones. This service aims to boost underbanked people's access to traditional banking services and broaden their financial inclusion. All Telecom Service Providers (TSPs) allow access to this service through the common code *99# on a mobile phone. The service offers an interactive menu for tasks including interbank account transfers, balance inquiries, mini statements, and other services. As of 30th November 2016, 50+ major banks and all GSM service providers offer this service, accessed in 12 languages, including English and Hindi. Various ecosystem partners, including Banks and Telecom Service Providers (TSPs), are brought together by this distinctive interoperable direct-to-consumer service [7].

AADHAR-ENABLED PAYMENT SYSTEM (AEPS)

AEPS is a banking concept that allows for electronic financial transactions to be made at a Point of Sale (PoS) or Micro ATM using Aadhaar authentication through the Business Correspondent (BC) or 'Bank Mitra' of any bank [7].

UNIFIED PAYMENTS INTERFACE (UPI)

The Unified Payments Interface(UPI) is a platform uniting several financial services, like fund transfers, merchant payments, and seamless routing, into a single application for any participating bank. With the UPI, a single app can access multiple bank accounts. Additionally, it makes "P2P" collection requests possible, which can be pre-planned and paid for later. Each bank provides its own UPI application across multiple platforms including Android, Windows, and iOS [7].

MOBILE WALLETS

A mobile wallet functions like a digital wallet that holds money. You can use it to transfer money online to the wallet or connect your credit or debit card account to a mobile wallet application. It enables purchases using a smartphone, tablet, or smartwatch for consumers instead of using a physical card. An individual's account must be connected to the digital wallet in order to add money to it. In addition to private businesses like Mobikwik, Freecharge, Paytm, Oxigen, mRuppee, JioMoney, Airtel Money, SBI Buddy, itz Cash, Citrus Pay, Vodafone M-Pesa, Axis Bank Lime, ICICI Pockets, and SpeedPay, several banks also provide their e-wallets [7].

POINT OF SALE

The place where goods or services are sold is referred to as the point of sale (PoS). This can happen on a broad scale, such in a market or shopping mall, or on a smaller size, like at a checkout desk in a store. In Banking terminology, the PoS uses a small device called 'PoS terminal' for swiping credit and debit cards for the customer to make purchases. This site is also referred to as "point of purchase" [7].

INTERNET BANKING

Online banking, often known as net banking, virtual banking, or e-banking, is a digital payment mode that enables customers of banks or other financial institutions to carry out a variety of financial transactions *via* the institution's website [7].

DIFFERENT TYPE OF FINANCIAL TRANSACTIONS - NATIONAL ELECTRONIC FUND TRANSFER (NEFT)

The National Electronic Funds Transfer (NEFT) is a payment system that enables people, businesses, and corporations to electronically transfer money between bank branches. With the help of this system, money can be sent from any bank branch to an individual, business, or corporation that has an account at another branch in the nation that is a part of the programme and participates in it. Individuals, businesses, or corporations with accounts at a bank branch may use NEFT. With a limit of Rs. 50,000 per transaction, people without bank accounts can also transfer money using NEFT by depositing cash at NEFTenabled branches with the instruction to do so. NEFT runs in hourly batches, with six settlements on Saturdays and twelve settlements during the week.

REAL TIME GROSS SETTLEMENT (RTGS)

Fund transfers using the RTGS payment system are continually and instantly paid on an individual basis without netting. "Real-time" refers to the instructions being processed right away, while "gross settlement" alludes to the settlement of each individual funds transfer order. Payments are final and cannot be cancelled because they are settled in the Reserve Bank of India's books. The RTGS system has a minimum remittance amount of 2 lakh and no maximum, and it is primarily designed for high-value transactions. RTGS timings are from 0900hrs to 1630hrs on weekdays and from 0900hrs to 1400hrs on Saturdays to settle customer transactions (Actual timings may change based on the bank branches and RBI guidelines) [7].

ELECTRONIC CLEARING SYSTEM (ECS)

The option to pay utility bills, such as mobile and energy bills, insurance premiums, card payments, loan repayments, *etc.*, is offered through ECS. Banks, businesses, corporations, government agencies, and other entities that collect payments will provide better customer service as a result of the elimination of the need to issue and manage paper-based payment instruments [7].

IMMEDIATE PAYMENT SERVICES (IMPS)

Mobile devices can access IMPS, a quick and always-available electronic fund transfer service that connects banks. Through mobile phones, the internet, and ATMs, it is a potent way to instantly transfer money within India. IMPS is a safe and economical choice that offers both monetary and non-monetary advantages [7].

MOBILE BANKING

Financial institutions, particularly banks, provide consumers the option of using their mobile phones to conduct a variety of financial activities. The financial institution must supply a software programme, sometimes known as an app, for this service. Every bank offers a mobile banking app made specifically for mobile operating systems including Android, Windows, and iOS [7].

MICRO ATMS

Millions of Business Correspondents (BC) use the Micro ATM to provide consumers with necessary financial services. The technology makes speedy transactions possible for Business Correspondents, who may be small-business owners who serve as "micro ATMs." This mini platform operates using inexpensive gadgets (micro ATMs) connected to national banks. This makes it possible for anyone, regardless of bank affiliation, to make a quick deposit or withdrawal of money. Every BC will have access to the device, which connects *via* a cell phone. Customers should only verify their identity before making a deposit or withdrawal from their bank accounts. The BC's cash drawer will be used to remove the funds. In essence, BCs will serve as consumers' banks, and all they have to do is, use the client's UID to confirm the client's identity. Basic transaction types like deposits, withdrawals, fund transfers, and balance inquiries will be supported by the micro ATM [7].

FACTORS THAT CONTINUE TO DRIVE DIGITAL PAYMENTS IN RURAL INDIA

Several mobile wallet and e-commerce businesses are focusing on rural areas and employing strategic tactics to capture a significant market share.

Recent events such as demonetization and its subsequent impacts have brought about significant changes in the market dynamics. Prior to demonetization, only a small number of merchants and customers from rural areas were utilizing digital payments. However, after de-monetization, there has been a tremendous increase in users utilizing digital payment interfaces. Other developments contributing to this transformation include the timed issuance of payment banks, mobile wallet transaction norms relaxation, and the improvement of internet connectivity in rural areas. The government has also encouraged the public to use UPI (United Payment Interface) through incentive schemes and has reduced service tax for digital payments for government services. These changes signify a major shift towards digital payments [8, 9].

Increasing Smartphone Penetration

In India, the rate of smartphone adoption has grown rapidly, from an estimated 54% in 2020 to a predicted 96% by 2040, when nearly everyone will own a smartphone. Since their inception, mobile payments have gained popularity among both urban and rural users. Fintech businesses are continually developing new products and services that facilitate payments in order to advance financial inclusion. Users in many locations are becoming more familiar with trends like Neo Banking, Fraud prevention, Buy Now Pay Later, QR code-embedded bill payments, and virtual wallets [10].

Digital Payments Replacing `Traditional Banking'

In rural India, the younger generation is accepting innovation spontaneously and fast. Digital payments have become more user-friendly and have eliminated the old problems of sending money through banks, which is what is driving this trend.

Even the smallest fund transfers used to be a laborious operation that involved gathering tokens, standing in lines, and interacting with cashiers. Those days are long gone [10].

Digital Payment Adoption for Rural Stores

The use of debit/credit cards, debit/credit cards with dynamic QR payments, inventory, e-point of sale, and neo-banking capabilities is all part of the key trend of digital payments on technology-powered offline retail apps [10].

Neo-banking alone raises the bar for rural digital adoption because the bulk of rural citizens work in B2C businesses with numerous layers of payment cycles to stakeholders. Group or individual payouts, a record of incoming and outgoing transactions, and, in certain circumstances, the opportunity to apply for a loan or purchase insurance are some of the neo-banking elements of a store's payment collecting app.

Prices are reasonable, and any required integrations may be completed quickly and easily. It is essential to have a stylish dashboard with alerts of actions taken on the mobile payment app [10 - 16].

Simplicity

Most mobile app users in rural areas need simplicity. The user interface of the software should be simple, clear, and intuitive. Although internally complex, the user should perceive the navigation, process, and concluding acknowledgements as straightforward and uncomplicated. The application should also be available in at least six to seven regional languages [10].

Speed

These tiny gaps between "send payment" and "paid" are considered significant and anxiety-inducing for humans. As per rural human mind psychology, the more time-consuming a digital payment is, the more likely it is to be untrustworthy. Nearly 70 percent of India's population lives in rural areas, where the majority of income is earned through labour, sweat, and blood, making problems like money loss panic-inducing. Rural business owners have greater security while moving swiftly [10].

An Edifying Campaign with a Focus on the Security of Digital Payments

With digital fraud-related news on the rise, awareness of digital payment security is essential. Fintech has built-in defenses for fraud detection and prevention, which makes merchants feel more secure while relying on digital payments.

Adopting digital technology brings many benefits and amazing capabilities, but it also comes with a certain fear factor. The convenience, ease of use, and speed offered by fintech solutions are designed so that "a known devil is better than an unknown angel".

To raise awareness of the transparency and trust that the digital world brings, the Ministry of Electronics and Information Technology has taken the necessary steps to set up a staffed digital financial hub, the Common Service Center (CSC). Additionally, they have the authority to address the concerns and questions of rural people and clarify that digital banking is subject to government regulation and that rural people have access to certain digital financial opportunities. The government has allocated over Rs. 65 crore for the same [10].

Fig. (**1**) shows a summary of digital payment transactions made over the course of the previous four financial years.

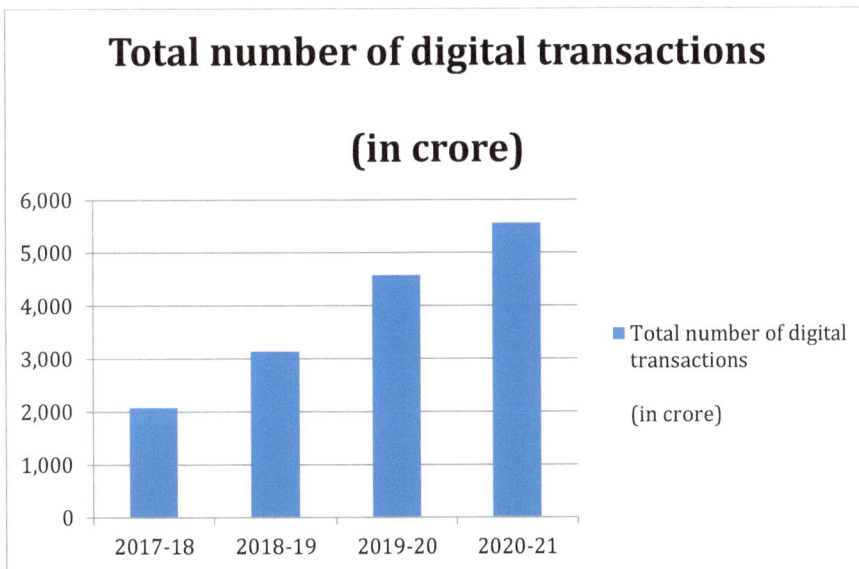

Fig. (1). A bar graph showing # number of digital transactions.

This bar graph shows the # number of digital transactions (in crore) in the financial years 2017–18, 2018–19, 2019–20, and 2020–21.

Note: Digital payment modes considered are BHIM-UPI, IMPS, NACH, AePS, N*ETC*, debit cards, credit cards, NEFT, RTGS, PPI and others. Source: RBI, NPCI and banks

Rising use of Digital Payments in Rural India

Digital payments have been increasingly popular in India's rural areas in recent years. More and more people are adopting digital payment services for their daily purchases as smartphone use and internet connectivity increase. The National Payments Corporation of India reported that compared to the prior year, there has been a 29% increase in digital transactions in rural areas during the financial year 2020–21.

The bar graph in Fig. (**2**) shows that the overall volume of digital payment transactions (in millions) in rural areas has been rising over time.

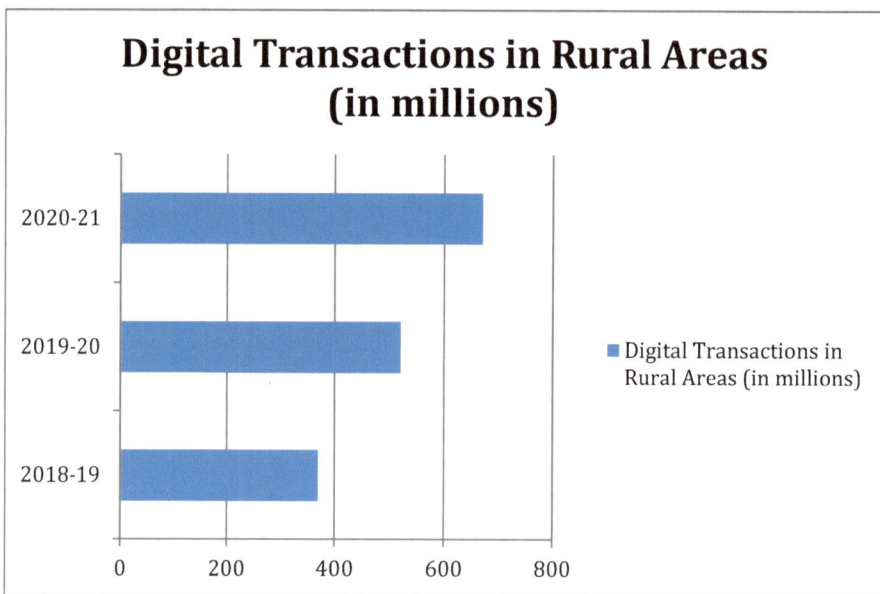

Fig. (2). A bar graph showing the total number of digital transactions in rural areas.

Around 370 million digital transactions took place in 2018–19; this number rose to 520 million in 2019–20; and then to 670 million in 2020–21. This demonstrates the steady expansion of digital transactions in India's rural areas.

As shown in Fig. (**3**), UPI (Unified Payments Interface) has a 42% market share among digital payment methods in rural India. AEPS (Aadhaar Enabled Payment System) and IMPS (Immediate Payment Service), which have respective market shares of 32% and 14%, come in second and third place to UPI. With a combined share of 12%, other digital payment methods include cards, mobile banking, and USSD (Unstructured Supplementary Service Data).

Digital payment modes - Rural India

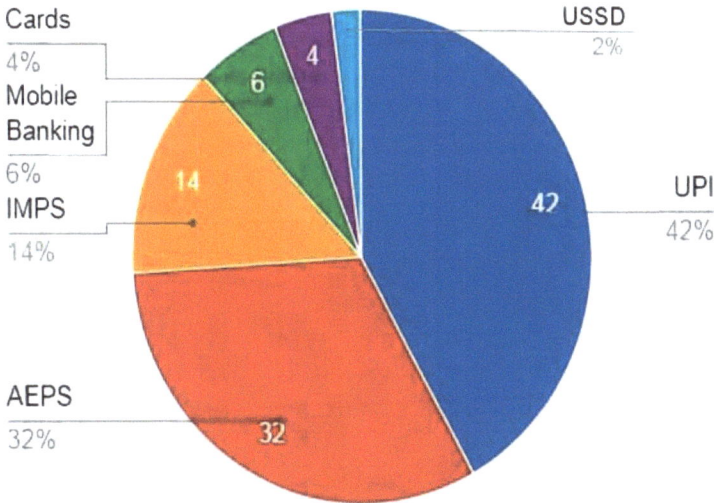

Fig. (3). Pie chart showing the usage of different modes of digital payments in rural India.

Convenience, safety, and financial inclusion are just a few advantages of the growing use of digital transactions in rural India. Digital payments have also contributed to a decline in cash use, which is frequently used for illegal operations and tax avoidance. The Indian government has been promoting digital payments as part of its Digital India project, which aims to convert the nation into a cashless society and knowledge economy that is empowered by technology.

CHALLENGES TO ADOPT DIGITAL PAYMENTS IN RURAL INDIA

India's rural economy forms a significant part of the country's economy. This is why many fintech companies, especially in rural India, are making financial inclusion a high priority. Before de-monetization, the majority of transactions in India were done in cash. However, India's urban areas are steadily going digital thanks to the government's brilliant initiatives.

Rural areas are underdeveloped due to a lack of modern schooling and a lack of digitization. Given that the majority of the population (65%) lives in rural areas, there is great potential to help India reach its goal of a $5 trillion economy by 2024-2025. The "digital literacy gap" is a huge hole that must be filled. The government launched programs like Digital India in 2015 with the aim of bridging this gap and transforming India into a digitally enabled nation [11, 12].

Trust Factor

Residents of remote areas find it difficult to establish confidence online. These people are technologically ignorant, thus they have no interest in learning how to utilize digital payments. Because they are unable to understand the advantages of disclosing their financial information to another person online, they are unable to build trust. They fear for their safety and think that anything concerning banking and money should be done with caution for security concerns [12].

Lack of Digital Literacy

Jio increased internet affordability and accessibility. The low-cost smartphone mania completely engulfed rural and lower middle-class populations. The citizens of India had regular access to these services. They were unable to understand the services and how to use them, thus it was useless [12].

In India, only 38% of households are computer literate. Rural communities are open to using Internet payment methods. However, a lack of digital literacy is a barrier [12].

The Comfort in Cash

The simplest form of payment for those who live in rural areas is cash. In India, cash is used in 72 percent of consumer transactions, according to a Credit Suisse analysis [12].

The Digital Infrastructure

In the modern world, a strong digital infrastructure is essential. The educational campaigns yield positive outcomes. Digital literacy is being passed down from person to person in India's rural communities. Rural populations continue to rely almost exclusively on cash since they lack the necessary infrastructure. For instance, tiny shops in rural areas lack card readers known as PoS terminals, a type of digital payment system [12].

RURAL INDIA'S BANK ACCOUNTS ARE EXPLOITED IN FINANCIAL CYBER FRAUD

According to a cyber unit probe, The Pradhan Mantri Jan DhanYojana (PMJDY) accounts are being sold to cyber criminals for INR 5000 per month. The cybercriminals are defrauding consumers of their money placed in various banks throughout the state utilizing bank accounts opened under PMJDY. By paying each account holder INR 5000 per month, the PMJDY accounts were rented out. The gang members even rented out some individual's credit cards for INR 8000.

This operational method has been adopted by a number of gangs that operate in Madhya Pradesh and Odisha.

The gang's method of operation is pretty straightforward. The poor and illiterate are approached by fraudsters to open Jan Dhan accounts so that they can benefit from government programmes. Thumb impressions made on plain papers were digitized for the police inquiry. Rubber thumbs are made by fraudsters using polymer chemical stamping equipment to maintain cash withdrawal regulations. Photoshop is also used to clean up photos.

Police investigations show that cybercrooks are buying bank accounts from villagers to make money to lure more people into committing fraud. This practice of "selling" bank accounts is widespread in areas such as Jamtara in Jharkhand. Delhi Police contacted the Ministry of Finance and RBI to address bank vulnerabilities during account holder verification and paperwork. A state-level coordinating committee meets quarterly in the presence of the Managing Director of the RBI to analyze gaps in the banking system. The RBI had previously warned that Jan Dhan's accounts were "highly vulnerable" to fraud, urging banks to be vigilant against such activity.

Criminals who gain unauthorized access to savings accounts may use third parties known as "money mules" to launder the proceeds of fraudulent schemes such as identity theft and phishing. Fully KYC-compliant Jan Dhan accounts have withdrawal limits set at INR 10,000 per month, however, for account holders who are not fully KYC compliant or partially KYC compliant, the monthly limit is INR 5,000. The monthly deposit limit is INR 1 lakh [13, 14].

CONCLUSION

The article discusses the challenges that arise due to the limited scope of opportunities in digital payment market dynamics. Although the use of digital payments can advance the processing of payments in rural areas, a significant portion of the population is still unfamiliar with the Indian financial system. A lack of digital literacy in rural areas is a major concern, especially since India is projected to have 1 billion smartphone users by 2026, which increases the risk of cybercrime. To bridge the digital divide, Raj institutions should be extended to rural areas and various government department officials, such as teachers, village functionaries and development officers, and healthcare professionals, should be trained in cyber security to prevent financial fraud. To support the development of digital payment dispensation in the rural sector, it is crucial to promote digital payments and explore new potential opportunities. Despite the social and technological limitations, the increasing popularity of mobile Internet banking in India may lead to a cashless economy.

FUTURE SCOPE

The future scale of UPI digital payments in rural India is huge and could bring financial inclusion to millions of people. A Boston Consulting Group (BCG) report shows that UPI transactions in rural areas are expected to grow at a compounded annual growth rate of 60-70% over the next 5 years, potentially leading to reduced and improved cash consumption. For that purpose, there is the Government Transparency Deal.

However, the growth of digital payments also brings with it the risk of financial cyber fraud. The Reserve Bank of India reported a 159% increase in cyber fraud incidents in the country's banking sector in 2020. This risk could be exacerbated in rural areas where many people may not be familiar with digital payment systems.

Addressing these challenges will require government and industry actors to work together to raise awareness, improve infrastructure, and ensure digital payment systems are safe and easy to use for all users. Biometrics such as fingerprint and facial recognition, and robust fraud detection and prevention systems can also help prevent fraud and secure transactions.

Despite these challenges, the potential benefits of UPI digital payments in rural India are enormous, including increased financial inclusion and reduced corruption. Govt. should work to build this into a trusted digital payment ecosystem [15, 16].

REFERENCES

[1]   P. Dhabhai, "The shift to digital payments: Empowering rural areas to make transactions seamless", *Times of India News,* vol. 29, 2021p. 3. Available from: https://timesofindia.indiatimes.com/blogs/voices/the-shift-to-digital-payments-empowering-rural-areas-to-make-transactions-seamless/

[2]   L. Malusare, "Digital Payments Methods in India: A study of Problems and Prospects", *International Journal of Scientific Research in Engineering and Management,* vol. 3, no. 8, pp. 2590-2594, 2019.

[3]   M. Haaris, "Over 44 cr accounts opened under PM Jan Dhan Yojana, women account for 55%", *Bus. Today (Norwich),* 2021p. 7.

[4]   Available from: https://pmjdy.gov.in/

[5]   Available from: https://www.ideasforindia.in/topics/governance/the-digital-dream-upskilling-i-dia-for-the-future.html

[6]   Osama Manzar, "Recipe for 100% digital literacy before 2021", *Mint,* 2012.

[7]   Digital Payments Methods. Available from: http://cashlessindia.gov.in/digital_payment_methods.html

[8]   D. Dennehy, and D. Sammon, "Trends in mobile payments research: A literature review", *Journal of Innovation Management,* vol. 3, no. 1, pp. 49-61, 2015.
[http://dx.doi.org/10.24840/2183-0606_003.001_0006]

[9]   Jonathan Dharmapalan, "EY-mobile-money-the-next-wave", *EYGM Limited,* pp. 20-40, 2014.

[10] K. Mohan, "Trends that rule digital payments in rural India", *Times of India,* p. 6, 2022.

[11] Rohit Kumar, "Why Bridging The Digital Gap In Rural India Is Extremely Important", *Inc42,* p. 8, 2021.

[12] "The complexity of payments in rural India", *CNBC,* vol. TV18, p. 16, 2021.

[13] Ramashankar, "Cyber frauds rent Jan Dhan accounts to dupe people in Bihar", *Times of India,* p.14, June 27, 2020.

[14] A. Kumar, "Money siphoned in cyber fraud finds way to Jan Dhan accounts: An investigation" India Today, p.5, 19, 2019.

[15] "The Road Ahead for UPI in India", Boston Consulting Group, 2020.

[16] "Trend and Progress of Banking in India", Reserve Bank of India. Annual Report 2021-2022.

CHAPTER 5

Artificial Intelligence Techniques based PID Controller for Speed Control of DC Motor

Rama Koteswara Rao Alla[1,*], **Neeli Manoj Venkata Sai**[1] and **Kandipati Rajani**[2]

[1] *Department of Electrical and Electronics Engineering, R.V.R. & J.C. College of Engineering, Guntur, Andhra Pradesh, India*

[2] *Department of Electrical and Electronics Engineering, Vignan's Lara Institute of Technology and Sciences, Guntur, Andhra Pradesh, India*

Abstract: DC motor demand is rising in the industrial sector due to its efficiency and in contrast to AC motors, a DC motor's momentum can be easily adjusted. For industrial uses, making a highly regulated motor is essential. DC motors need to have excellent speed tracing and load regulation in order to operate satisfactorily. The speed of a DC motor was controlled in this work using proportional integral derivative (PID) controllers. This study used MATLAB to determine how a Proportional-Integral-Derivative (PID) controller affected the performance of a DC motor of the industrial type by selection of PID controller parameters using Zeigler's Nichols (ZN), Genetic Algorithm (GA), and Fuzzy Inference System. Nonlinearities and model uncertainties must be included in the control design in order to provide effective and efficient control. The higher-order systems could use the suggested strategies. The PID controller's primary function is to regulate motor speed based on incoming system data and auto-tuning. The findings of the simulation also demonstrate improved motor performance, which decreases rise time, steady state error, and overshoot, and increases system stability.

Keywords: DC motor, Fuzzy inference system, Genetic algorithm, PID, ZN method.

INTRODUCTION

PID controllers are the most typical and commonly used controllers in the industrial sector because of their straight forward design, safety, and dependability. Despite having greater maintenance costs than induction motors, DC motors have been employed in the industry for some time. High-speed contro-

---

[*] **Corresponding author Rama Koteswara Rao Alla:** Department of Electrical and Electronics Engineering, R.V.R. & J.C. College of Engineering, Guntur, Andhra Pradesh, India; E-mail: ramnitkkr@gmail.com

Neha Kishore, Pankaj Nanglia, Shilpa Gupta & Ashutosh Kumar Dubey (Eds.)

llability, steady-state and transient-state stability, and favorable torque-speed characteristics are needed for DC motors.

Recent advances in science and technology have made it possible to use high-performance DC motor drives in a variety of settings, including rolling mills, chemical processing, electric trains, robotic manipulators, and domestic appliances. They demand controllers to complete duties [1].

As a result, a control system uses a genetic algorithm that takes the system's effectiveness into account. Based on the concepts of genetics and natural selection, the genetic algorithm has been developed. A stochastic global search technique called genetic algorithms (GA's) imitates the course of natural development. The most effective controller for the system will always be considered when using genetic algorithms to tune the controller. The purpose of this project is to demonstrate how optimization can be achieved by using the GA method and FIS to tune a system. There is a method to make technologies more intelligent to work like humans which is the Fuzzy Logic [2].

Real-time parameter optimization is required to choose the optimal K_p, K_i, and K_d values to satisfy the requirements of users in a specific process plant. Finding the ideal values of K_p, K_i, and K_d can be done in a variety of ways. Genetic Algorithm, a technique for computation, is one of the methods. The principles of natural evolution are the foundation for search and optimization algorithms known as genetic algorithms [3]. This study provides a brief overview of PIDs, their tuning techniques, including Ziegler's Nichols, Genetic Algorithm and Fuzzy Inference System as well as DC motor control and their outcomes.

DC MOTOR MODELLING

An inertia model, which is the mechanical analogue of a DC motor, may be created by employing the control systems. Using interacting magnetic fields, an electric motor transforms electrical energy into mechanical energy. Resistance (R) and Inductance (L) are the mechanical components, whereas inertia constants (K_t, K_e, K_n), load inertia (J), damping (b), and angular position are electrical components. A further significant non-linear characteristic of DC motors is the saturation effect for output speed [4]. A mathematical type of transfer function is shown below.

$$\frac{\theta(S)}{V(S)} = \frac{K_t}{S[(LS+R)(JS+b)+K_tK_e]+K_nK_t} \qquad (1)$$

The equation above is the generalized transfer function of a DC motor. The transfer function of the DC Motor system is found with the relevant parameters in the generalized transfer function.

PROPORTIONAL INTEGRAL DERIVATIVE CONTROLLER

On the market since 1939, a proportional integral derivative controller is simple to operate and is still the most popular controller for feedback control of industrial sector activities. Proportional, Integral, and Derivative Control are the three words that make up the acronym PID controller. These three controllers work together to provide a control strategy for process control as shown in Fig. (**1**). Pressure, speed, temperature, flow, and other process variables are controlled using proportional integral derivative controllers [5]. There are techniques for obtaining the finest and most ideal values of K_p, K_i, and K_d. The engineers from 1950 invented the Ziegler and Nichols tuning approach, which established tuning principles to find and select the appropriate settings of PID controllers. Given is the equation for the PID controller.

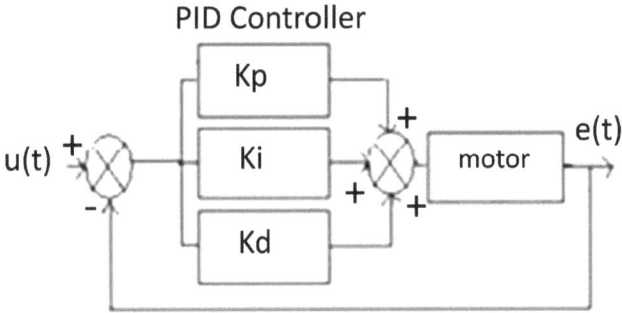

Fig. (1). Block diagram for PID controller.

TUNING METHODS

The PID Controller may be tuned using the techniques:

- Zeigler Nichols Method
- Genetic Algorithm Method
- Fuzzy Inference System (FIS)

Ziegler Nichols Tuning Method

Ziegler-Nichols tuning rules refer to two techniques for figuring out PID controller settings as shown in Table **1**. However, the most generally used technique for fine-tuning the PID controller is a simple one. Set the controller to P mode alone initially. The controller's gain (K_p) should then be adjusted to a low value. If it is low, the reaction ought to be slow. When the answer starts to oscillate, it increases by a factor of two and continues increasing. Lastly, tweak K_p until oscillations that are continuous are produced [6]. The ultimate gain (K_u) is what is meant by this. Keep in mind that the oscillations' time is referred to as the "ultimate period" (T_u).

The following list of stages constitutes the method:

The integral and derivative coefficients' profits must both be zero. The proportional coefficient is increased gradually (beginning at zero) until the system barely starts to oscillate regularly (sustained oscillation). The final benefit is now represented by the proportionate coefficient (K_u). Moreover, this oscillation phase is referred to as the last period (T_u). The Ziegler Nichols Tuning rule is then obtained from the following Table **1**.

Table 1. Ziegler nichols tuning rule.

Controller Type	Kp	K_i	K_d
PID	$K_u/1.7$	$T_u/2$	$T_u/8$

Genetic Algorithm Method

The use of a genetic algorithm has been acknowledged as an efficient and successful method for solving optimization issues. GA shown in Fig. (**2**) begins with an initial population of high-quality individuals, each of whom offers a potential solution to the issue at present. A fitness function evaluates performance. Each individual is basically subjected to three fundamental genetic operators of selection, crossover, and mutation [7].

By the use of these three essential techniques, new individuals who may outperform their parents can grow. After a certain number of generations, the algorithm stops when it encounters the people who are the best solution to the problem. Fig. (**2**) illustrates the process of the genetic algorithm.

There is a population of points produced after each repetition. The optimal resolution is getting closer to the population's best point and it generally, converges after numerous functional evaluations [8]. These three essential techniques facilitate the growth of new individuals who may exceed their parents. This method iterates through several generations until stopping when it identifies the person who best represents the problem's ideal solution. The flowchart in Fig. (**2**) above demonstrates how to implement the GA algorithm to optimize the PID controller's parameters for the specified system [13, 14].

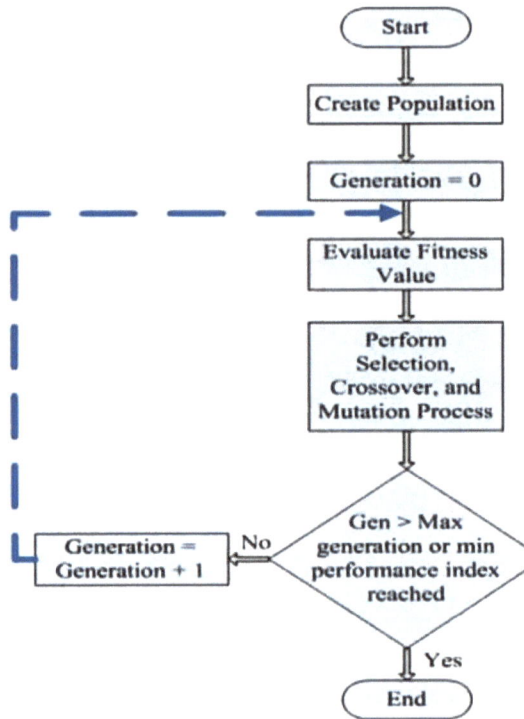

Fig. (2). Genetic algorithm.

Fuzzy Inference System Method

DC motor's speed may be managed using a variety of techniques, including PID control, fuzzy logic control, and neural network approach. The control technique is given a human-like intuition by the fuzzy method, which is also self-tolerant of less exact inputs. Technology may be made to be more intelligent to mimic humans through fuzzy logic. It, although considerably quicker, simulates how a human would make a judgment [9].

The Fuzzy Logic Controller (FLC) shown in Fig. (**3**) offers an algorithm that may transform a control strategy's inputs through three fundamental stages fuzzification, decision-making, and defuzzification before producing the output. With the use of specified membership functions, the inputs are transformed into linguistic variables. When seen from this angle, a collection of linguistic control rules connected to fuzzy implication and its rule of inference serve as the fundamental component of the fuzzy logic controller (FLC) [10]. For both inputs and outputs, the FLC-based Mamdani fuzzy inference system makes use of linear membership functions.

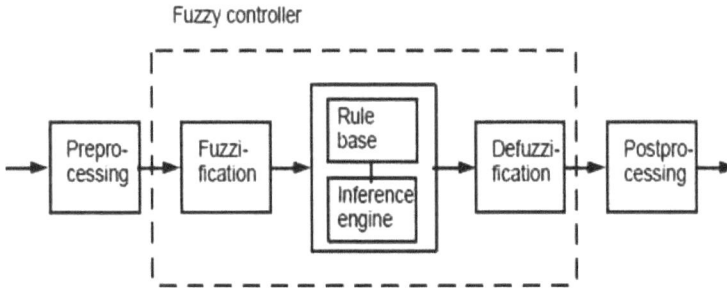

Fig. (3). Structure of fuzzy logic controller.

A collection of input membership functions, a rule-based controller, and a defuzzification procedure make up the fundamental components of a fuzzy logic control system. Fig. (1) shows the general layout of a fuzzy logic controller [11] that uses suitable membership functions to build a set of if-then rules that yield the specified input-output pairs created using MATLAB [12]. The number of fuzzy sets that make up linguistic variables, the mapping of measurements into support sets, the control protocol that defines controller behaviour, and the form of the membership functions are the four important aspects that must be chosen in order to execute the FLC. The linguistic variables used as inputs and outputs in the fuzzy logic controller design were divided into three categories: negative (N), zero (Z), and positive (P). The Fuzzy Inference System's rules are liste Table d in Table **2**. Three membership functions are present in the fuzzy input variable e(t), three membership functions are present in the fuzzy input variable ce(t), and three membership functions are present in the output variable.

Table 2. Fuzzy Rule Table.

ce(t) e(t)	N	Z	P
N	N	N	Z
Z	N	Z	P
P	Z	P	P

The triangular form was one of the two MFs that were used to model the FLC [15 - 18]. The above Table **2** represents the rules as they were translated from the fuzzy logic toolkit [19]. Moreover, FIS used membership functions to translate provided inputs into outputs using fuzzy logic rules, and the system was designed in MATLAB.

SIMULATION RESULTS AND DISCUSSIONS

MATLAB SIMULINK is used for simulation. The investigation demonstrates the best outcomes for speed control of DC motors using artificial intelligence approaches. Here, a pulse generator is used as an input, and the result is a DC motor's step reaction.

Using Ziegler Nichols Method

The PID controller is illustrated in Fig. (7), with K_p = 49.41, K_i = 0.075, and K_d = 0.01877. The time domain parameters are given in the table to demonstrate the system's performance (3). The critical time T_u = 0.15 sec. and the critical gain K_u = 84.

From Fig. (4), the peak time is around 2.48 seconds, and the settling time is approximately 2.4 seconds. The system has not been adjusted to its best and optimum solution, according to the above result.

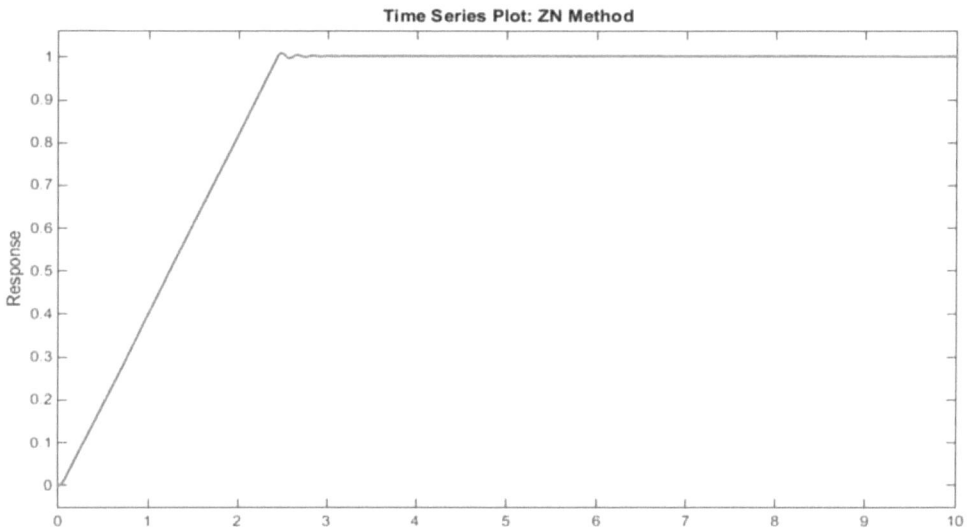

Fig. (4). Step response for ziegler nichols method.

Using Genetic Algorithm Method

By comparing several potential solutions, the Genetic Algorithm provides the most optimum option. The output response of the system with the use of the genetic algorithm method is shown in Fig. (5), when K_p = 19.88, K_i = 0.1376, and K_d = 0.5578.

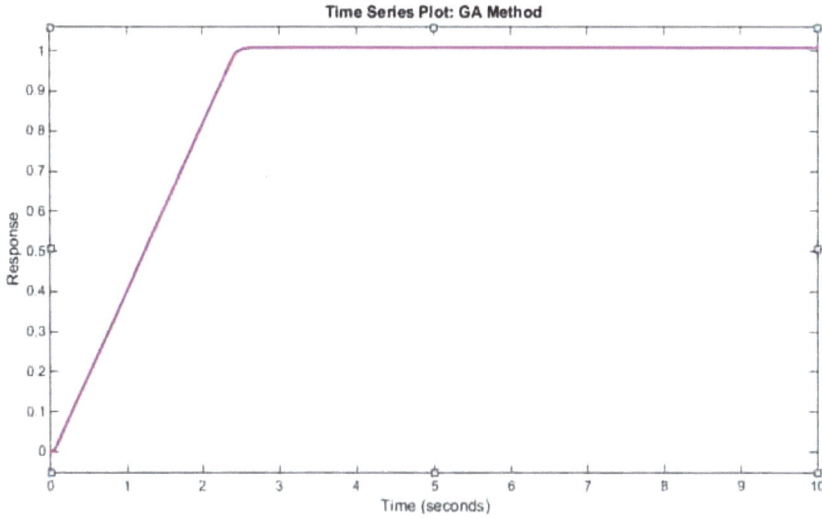

Fig. (5). Step response for genetic algorithm method.

From the above response, it can be observed that the peak time is about 4.92 sec and the settling time is about 2.3 sec. and the system gives optimum and comes to a stable state in less time.

Using a Fuzzy Inference System

The performance of the Fuzzy Logic controller may be considerably enhanced by changing the membership function's forms. Following FLC model adjustments and rules, the inputs and output MFs, as well as the Simulink Model of the Fuzzy Inference System, the response for the Fuzzy Inference System is shown in Fig. (6).

From the above response, it can be observed that the peak time is about 0.16 sec and the settling time is about 0.91 sec. There is an improvement in the response, and a reduction in the peak time and settling time of the system.

Fig. (7) presents the comparison between all three methods. From the above figure, it can be observed that the rise time, settling time, and the peak time of the system are calculated to show the performance of the controller system. Time domain specifications are given in Table 3. Output responses of the systems were analyzed and hence it was found that the Fuzzy Inference System is the better controller among the other controllers and the settling time for the system with Fuzzy Inference System has a faster speed of response than the other tuning methods.

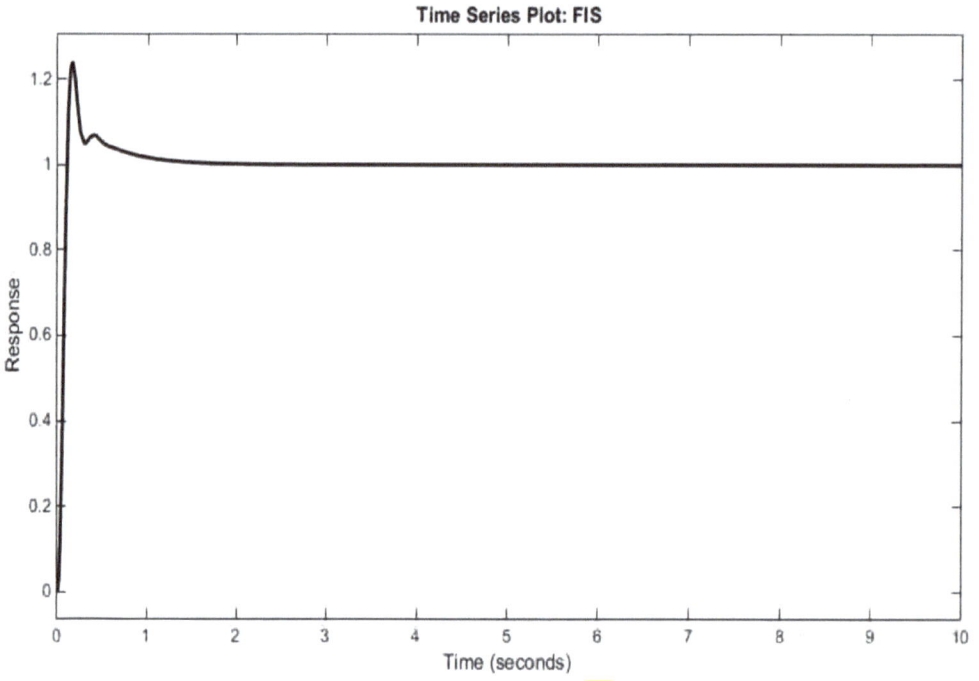

Fig. (6). Step response for fuzzy inference system.

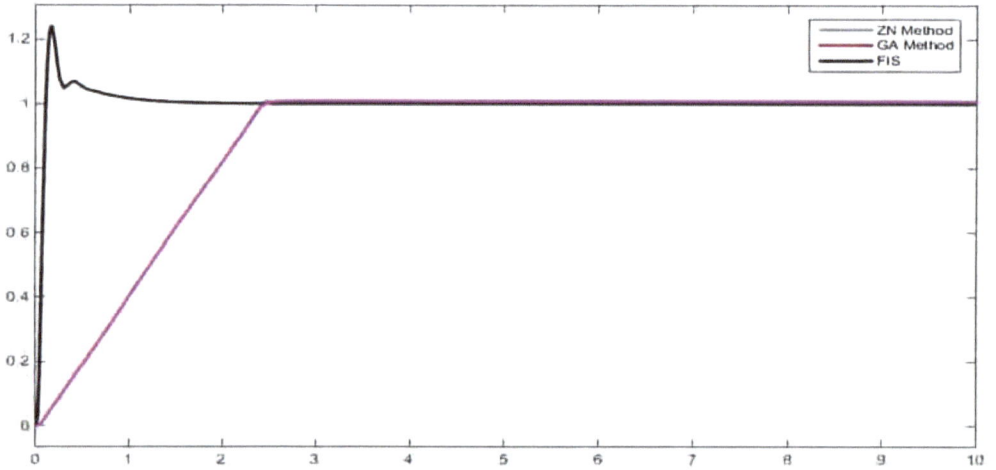

Fig. (7). Comparison between ZN, GA, and FIS responses.

Table 3. Time domain comparison between ZN, GA, and FIS responses.

Specifications	ZN Method	GA Method	FIS
Settling Time	2.4	2.3	0.9
Rise Time	1.9236	1.9370	0.0645
Peak Time	2.4884	3.0246	0.1659

CONCLUSION AND FUTURE SCOPE

This work considers the performance analysis of PID controller tuned by artificial intelligent techniques for speed control of DC motor. Three methods are considered. From the results, it can be concluded that the design of Fuzzy Inference System has much faster speed of response than Zieglers Nichols and Genetic Algorithm. The simulation result shows that the desired speed control of the DC Motor system may be accomplished using a Fuzzy Inference System. However, compared to the other approaches, Fuzzy Inference System performs significantly better in terms of peak time, rising time, and settling time. Finally, compared to the error computed using the conventional method, the error associated with the Fuzzy Inference System is significantly low. By examining the results made in this study, it can be concluded that the Fuzzy Inference System is a better method than the mentioned methods for properly controlling the speed of a D.C. motor. The work can be extended by considering hybrid techniques for PID tuning such as neuro-fuzzy, *etc.*

REFERENCES

[1]   R.P. Suradkar, and A.G. Thosar, "Enhancing the Performance of DC Motor Speed Control Using Fuzzy Logic", *Int. J. Eng. Res. Technol. (Ahmedabad),* vol. 1, no. 8, pp. 1-8, 2012.

[2]   R.R. Alla, J.S. Lather, and G.L. Pahuja, "PI Controller Performance Analysis Using Lambert W Function Approach for First Order Systems with Time Delay", *International Journal of Advanced Science and Technology,* vol. 86, pp. 1-8, 2016.
      [http://dx.doi.org/10.14257/ijast.2016.86.01]

[3]   Z.L. Gaing, "A Particle Swarm Optimization Approach for Optimum Design of PID Controller in AVR System", *IEEE Trans. Energ. Convers.,* vol. 19, no. 2, 2004.

[4]   S. Padhee, A. Gautam, Y. Singh, and G. Kaur, "A novel evolutionary tuning method for fractional order PID controller", *International Journal of Soft Computing and Engineering,* vol. 1, no. 3, 2011.

[5]   K. Rajani, "Modelling and control of switched reluctance motor", *BSc Thesis, Vignan's Engineering College.* Vadlamudi, Guntur, 2011.

[6]   K. Rajani, K. Rachananjali, K. Krishna, and V.S.L. Tirumala, "Speed control of 1-φ induction motor using 1-φ matrix converter", *International Journal of Control Theory and Applications,* vol. 9, pp. 267-271, 2016.

[7]   R. Singhal, S. Padhee, and G. Kaur, "Design of Fractional Order PID Controller for Speed Control of DC Motor", *International Journal of Sci-entific and Research Publications,* vol. 2, no. 6, 2012.

[8] N. Lekyasri, K. Rajani, "Lekyasri N, Rajani K.: PID Control Design for Second Order Systems, I.J", *Engineering and Manufacturing,* vol. 4, pp. 45-56, 2019.

[9] R.K.R. Alla, J.S. Lather, and G.L. Pahuja, "Comparison of PI controller performance for first order systems with time delay", *Journal of Engineering Science and Technology,* vol. 12, no. 4, pp. 1081-1091, 2017.

[10] S.N.V. Akhila, and K. Rajani, "Control of SRM using 3-level Neutral point diode clamped converter with PI and Fuzzy controller. International Advanced Research Journal in Science", *Engineering and Technology,* vol. 3, no. 8, pp. 211-217, 2016.

[11] S. Vaneshani, and H. Jazayeri-Rad, "Optimized fuzzy control by particle swarm optimization technique for control of CSTR", *World Acad. Sci. Eng. Technol.,* p. 59, 2011.

[12] K. Sundaravadivu, and K. Saravanan, "Design of fractional order pid controller for liquid level control of spherical tank", *Eur. J. Sci. Res.,* vol. 84, no. 3, pp. 345-353, 2012.

[13] A. Varsek, T. Urbancic, and B. Filipic, "Genetic algorithms in controller design and tuning", *IEEE Trans. Syst. Man Cybern.,* vol. 23, no. 5, pp. 1330-1339, 1993.
[http://dx.doi.org/10.1109/21.260663]

[14] E. David, "Goldberg: Genetic algorithms in search, optimization and machine learning. The university of alabama", *Addison-Wesley Publishing Company Inc,* 1989.

[15] O. Dwyer, "PI and PID controller tuning rules for time delay process: A summary", *Proceedings of Irish Signals and Systems Conference,* pp. 589-592, 1999.

[16] Chegudi Ranga Rao, Ramadoss Balamurugan, RamaKoteswara Rao Alla, "Artificial rabbits optimization based optimal allocation of solar photovoltaic systems and passive power filters in radial distribution network for power quality improvement", *Int. J. Inte. Engin. & Sys.,* vol. 16, no. 1, pp. 100-109, 2023.

[17] R. Matusu, "Application of fractional order calculus to control theory", *International Journal of Mathematical Models and Methods in Applied Sciences,* vol. 5, no. 7, pp. 1162-1169, 2011.

[18] RamaKoteswara Rao Alla, Kandipati Rajani, Ravindranath Tagore Yadlapalli, "Design of FOPID controller for higher order MIMO systems using model order reduction", *Int. J. Sys. Assu. Engin. & Manag.,* vol. 14, no. 5, pp. 1660-1670, 2011.

[19] W.M. Elsrogy, M.A. Fkirin, and M.A.M. Hassan, "Speed control of DC motor using PID controller based on artificial intelligence techniques", *International Conference on Control, Decision and Information Technologies (CoDIT),* pp. 196-201, 2013.
[http://dx.doi.org/10.1109/CoDIT.2013.6689543]

CHAPTER 6

Recognition of Diabetic Retina Patterns using Machine Learning

Parul Chhabra[1,*] and **Pradeep Kumar Bhatia**[1]

[1] *Department of Computer Science and Engineering, G. J. University of Science and Technology, Hisar, Haryana, India*

Abstract: Medical images contain data related to the diseases and it should be interpreted accurately. However, its visual interpretation is quite complex/time-consuming and only medical experts can examine this data precisely. In case of diabetes, the retina may be damaged and it is quite complex to examine its impact on the retina because there are a lot of vessels inside the human eyes that may be changed due to this disease and manual interpretation of these changes consumes excessive time. In order to overcome this issue, in this paper, a contour-based pattern recognition method (CBPR) is introduced that can recognize multiple patterns in sample retina images. Comparative analysis with the segmentation-based method (SBPR) shows that it outperforms in terms of performance parameters (*i.e.* Accuracy/Sensitivity/Specificity *etc.*).

Keywords: Classification, Machine learning, Medical image analysis, Pattern recognition.

INTRODUCTION

A medical image contains data about the disease/injury that can be recognized by some visual key points, called patterns. Manual interpretation of these patterns may be an error-prone and a tedious task but these days, computer vision and machine learning algorithms can be utilized to analyze these patterns in medical images, termed as pattern recognition process. Patterns may vary with respect to each disease type as shown in Figs. (**1** to **4**).

---

* **Corresponding author Parul Chhabra:** Department of Computer Science and Engineering, G. J. University of Science and Technology, Hisar, Haryana, India; E-mail: parul15march@gmail.com

Neha Kishore, Pankaj Nanglia, Shilpa Gupta & Ashutosh Kumar Dubey (Eds.)

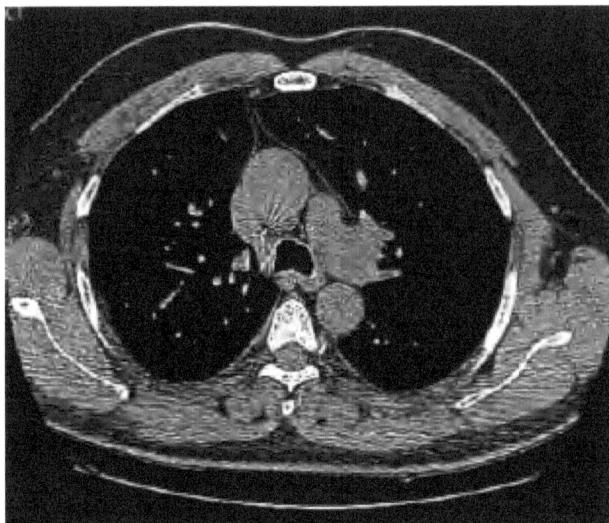

Fig. (1). CT scan image of lungs [2].

Fig. (**1**) shows the CT scan of the lungs having different patterns with respect to the shape of the lungs.

Fig. (**2**) shows the lighted tumor pattern in the brain MRI sample image.

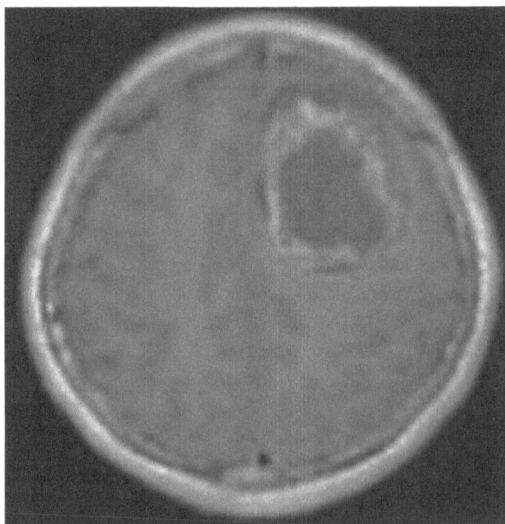

Fig. (2). Tumor pattern in brain MRI image [2].

Fig. (**3**) shows the patterns in knee MRI image. It can be used to find out the damage nerves or broken bones.

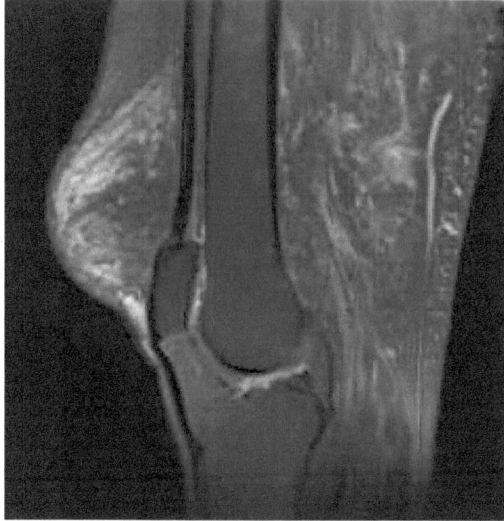

Fig. (3). Patterns in knee MRI image [3].

Fig. (**4**) shows pattern formation in the diabetic retina. This sample image can be compared with a healthy retina, for decision making.

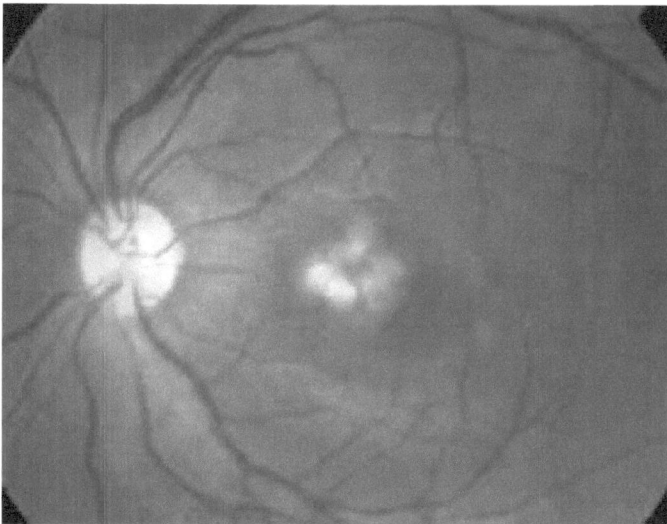

Fig. (4). Pattern formation in diabetic retina [4].

Fig. (**5**) shows the different ECG patterns of the heart. In case of medical emergency, it can be used for quick diagnosis of patients.

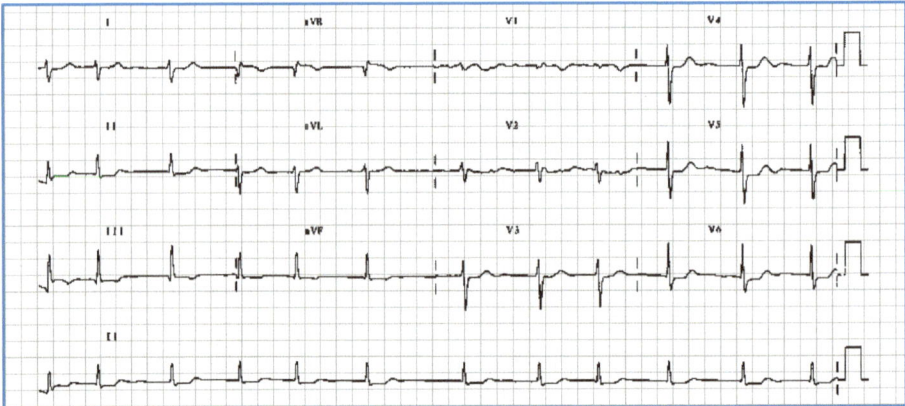

Fig. (5). ECG patterns of heart [5].

Limitations of pattern recognition using the machine learning approach:

• The accuracy of the training/testing model highly depends on input samples.

• Incorrect data labeling may lead to error-prone outcomes.

• Different patterns are associated with various disease types, so a single pattern recognition algorithm cannot be used for all disease types.

• There is a need to validate pattern recognition datasets by medical experts, otherwise it may lead to an incorrect diagnosis.

Fig. (**6**) shows the application of pattern recognition for medical imagery as explained below.

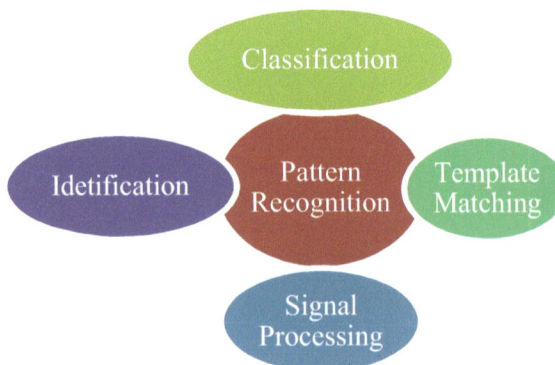

Fig. (6). Pattern recognition for medical imagery.

Following are the different applications of pattern recognition:

(a) Classification: It can be used to identify a specific disease type by comparing input samples.

(b) Identification: It can be used to identify different features in given samples.

(c) Template Matching: It can also be used to match templates in the given input.

(d) Signal processing: Different medical devices produce various signal types and they can be used to interpret these signals [37 - 39, 41].

LITERATURE SURVEY

S. F. Ershad *et al.* [6] introduced a deep learning-based genetic algorithm to diagnose cervical cancer. It extracts texture data to form local ternary patterns and finally, the multi-layer classifier is used for disease detection. Analysis indicates its performance in terms of higher detection rate and it can be extended for computer vision applications.

Z. Li *et al.* [7] developed a based scheme to identify Parkinson's disease. It processes input handwriting samples to produce patterns and uses a continuous convolution network for pattern classification to identify disease. Experimental results show that it can efficiently classify the handwriting attributes with respect to . disease recognition as compared to the existing method.

AP. B. Budiarsa *et al.* [8] used swarm optimization with extreme learning to detect the myoelectric patterns. It uses a hybrid swarm wavelet with neural network hidden layers to build a training model that is used for pattern recognition. Experimental results show that the proposed scheme has higher accuracy in recognizing input patterns as compared to the classical swarm method.

H. Zhang *et al.* [9] introduced a swarm intelligence based method to recognize patterns produced by the movement of muscles. It extracts features from input samples using a swarm algorithm and performs classification to recognize the movement of limbs. Comparison with traditional algorithms (whale optimization/Particle Swarm/sparrow search) shows that the genetic algorithm has higher detection accuracy.

A. Subasi *et al.* [10] introduced a deep learning-based disease diagnosis scheme that performs deep feature extraction over input images and finally, pattern recognition is used to detect the disease and a diagnosis plan is generated by transfer learning models. Experimental results show that it can predict the high-risk patients with optimal false negative rates.

D. M. Radha Devi *et al.* [11] developed a technique to keep a record of the health status and history-related information of patients. It uses a local binary pattern along with facial recognition to store/manage/*fetch* medical data. Experimental results show that it can efficiently maintain healthcare records and can be integrated with real-time healthcare services.

J. A. Patel *et al.* [12] investigated the role of machine learning in the recognition of lung diseases. A study found that the types of lung diseases can be classified on the basis of sound signals being produced by the lungs. The analysis also shows that different training/transfer learning models can be defined to increase the accuracy of disease recognition.

T. Tuncer *et al.* [13] used the combination of discrete wavelet concatenated Mesh tree and ternary chess pattern for the recognition of Electrocardiogram (ECG) signals. It extracts different features from input ECG signals using a ternary chess pattern and then features are selected through neighborhood component analysis. Finally, detection accuracy is ensured using k-nearest neighbor/support vector classifiers. Analysis shows that it offers a higher detection/recognition rate as compared to existing solutions.

S. Qiao *et al.* [14] developed an edge extraction method to process medical image data. It uses an improved local binary pattern method to extract the circular edges from images. It also performs noise and edge filtering to improve the contrast ratio. Analysis shows that different file formats (X-ray/CT/MRI) can be processed using this scheme with higher accuracy as compared to existing methods.

U. Ogiela *et al.* [15] investigated the relation of visual pattern recognition with the different parameters *i.e.* expected values, existing knowledge, and perception thresholds. A study shows that contents of the image dataset can be recognized accurately using all above parameters and a predictive algorithm can be developed to reconstruct the patterns for input images.

K. Yuki *et al.* [16] proposed a solution to analyze the various types of sepsis disease (bacterial/fungal/viral) using a pattern recognition method. Each type of sepsis has a different pattern for infected cells and receptors are used to distinguish between healthy and infected cells with respect to the disease type. Experimental results show that it can efficiently detect patterns and practitioners can recommend the treatment as per the infection level.

A. I. Korda *et al.* [17] introduced a pattern recognition scheme for the treatment of psychotic disorders. Analysis shows that it can classify the disordered patterns before/after the antipsychotic treatment and practitioners can improve the treatment plan as per health recovery rate. It can be further extended to recognize different diseases.

K. Kitagawa *et al.* [18] investigated health issues related to caregivers in hospitals and found that different activities were involved in handling patients with a common problem of back pain. The authors introduced a pattern recognition algorithm that can recognize the body motions during patient handling and can detect motion patterns that can cause back pain. Its performance was compared with different algorithms and analysis shows that it has the highest accuracy with random forest as compared to others (support vector machine/decision tree/neural network/ k-nearest neighbor/naive Bayes/logistic).

H. Gunasinghe *et al.* [19] developed a pattern-based classification method to identify glaucoma disease. It uses random forest/logistic regression classifiers for feature extraction from input samples and finally, cross-validation is performed using ResNet architecture. Analysis shows its performance in terms of disease detection but at least two different feature sets are essential to produce the desired results.

O. Leombruni *et al.* [20] have developed a pattern matching solution for the medical data acquisition using magnetic resonance. First of all, input samples are processed at low resolution and its output is further refined using high resolution and finally, magnetic resonance fingerprints are reconstructed. Analysis shows that it offers higher accuracy/sensitivity for image reconstruction as compared to traditional method.

R. Ahsan *et al.* [21] presented a convolution deep learning neural network algorithm for image recognition that assigns attribute weighting models to classify structural variances at raw genomic/proteomic series for different patients. Analysis shows that it delivers optimal disease prediction accuracy. However, the efficiency of the pattern recognition process can be further improved with the combination of external attributes (weight/feature/region).

D. G. Andrés *et al.* [22] have proposed a machine learning-based pattern recognition scheme for the detection of muscle diseases. MRI fingerprints/heat map patterns are used to distinguish between neuromuscular and on-set disorders. Experimental analysis shows it is suitable for muscular imaging in a real-time environment.

Y. Ma *et al.* [23] have developed a prediction model for disease detection. It uses a neural network to build a prediction model by examining the differences between the collected samples of recovered/infected patients. Experimental results show that its prediction accuracy depends on the input sample size as well as compared to the traditional neural network.

L. Borne *et al.* [24] have presented an automated classification approach based on pattern recognition for the detection of psychiatric illnesses. It uses different classifiers *i.e.* support vector classifier/patch estimator /Convolution Neural Network (CNN) over sample input. Analysis shows that it can accurately identify the multiple patterns using a patch estimator as compared to other methods.

F. Loncaric *et al.* [25] have proposed a machine-learning approach to analyze the echocardiographic data. During the cardiac cycle, it automatically recognizes different patterns for the identification of functional phenotypes. Analysis shows that it can easily distinguish between different patterns (blood cells/deformation profiles/tissues) efficiently and can be used to interpret medical data at a large scale with higher accuracy.

C. Djellali *et al.* [26] proposed a deep learning-based pattern recognition method for the diagnosis of breast cancer. It performs multiple steps *i.e.* feature identification, selection and extraction. over detected patterns. Analysis shows that the use of feature-based training models reduces the computational cost and offers higher accuracy for disease detection.

S. A. V. Begum *et al.* [27] have proposed a feature extraction and pattern recognition scheme for the detection of neurodisorders. It extracts the features using recurrence quantification and fast walsh-hadamard Transform to relate the gait patterns with respect to patients. Different classifiers (SVM/KNN/random forest/Discriminant Analysis) are used for the classification of patterns and analysis shows that only Discriminant Analysis has the highest detection accuracy as compared to others.

I. Urbaniak *et al.* [28] investigated the relationship between image compression algorithms (used to process medical images) and the accuracy of pattern recognition. The analysis found that resolution and image quality may be degraded by compression algorithms and thus can degrade the output of pattern recognition methods. However, results can be improved using the JPEG standard as compared to the JPEG2000 standard for medical images by varying compression ratios.

M. Liu *et al.* [29] presented a machine learning-based solution to process the ECG data. It uses semi-supervised to build the recognition model using CNN classifiers. Using the Laplacian matrix, it constructs a revisable prediction model to reconstruct the sample data. Analysis shows that it can extract different features (*i.e.* spectral/temporal) from EEG data with higher accuracy as tested over publically available ECG datasets.

Y. Jiang *et al.* [30] have presented a machine-learning approach to the diagnosis of shoulder disorders. It uses a convolutional neural network (CNN) based classifier to compare electromyography patterns with other patterns generated by muscular motions (resting/drinking, backward and forward motion/abduction, *etc.*). Analysis shows that its accuracy of pattern recognition varies with respect to . different motions and speeds. However, it can be improved by adding different motion parameters in the training model.

P. Feng *et al.* [31] used a random forest-based scheme to recognize protein sequences by extracting the optimal features from the given dataset. Analysis has shown that its performance depends on the population of selected features and it is still an open issue. However, cross-validation test ensures its accuracy up to the satisfactory level.

Y. J. C-Pino *et al.* [32] investigated the gait patterns of healthy people and patients suffering from disease and compared the data using wavelet and peak detection methods and analysis found that this method can distinguish between the patterns of healthy and patient's samples accurately and it can be further extended for the analysis of different parameters (*i.e.* motion/speed/arm swings/ankle movements *etc.*)

K. Vatanparvar *et al.* [33] have developed a method for automated cough detection. It builds a Gaussian mixture model using audio patterns collected from different patients and uses a neural networks-based cough model to relate the output with respect to patients. Analysis has shown that it can recognize only a few types of coughs and its accuracy also depends on the density of the dataset used for examination.

J. Waldthaler *et al.* [34] developed a method to recognize facial expressions of patients having Parkinson's disease. It builds a model using different parameters (happiness/ surprise/disgust/anger/ fear/ sadness /neutral face, *etc.*). Analysis of visual patterns shows that the proposed scheme can identify different emotions on the basis of patterns but its accuracy depends on the size of the input dataset.

L. Bai *et al.* [35] introduced a pattern recognition receptor scheme for cancer diagnosis. Receptors are used to recognize the patterns of cancer cells/healthy immune cells (during diagnosis). Analysis shows that, after clinical trials, diagnosis and treatment plans can be updated and it can be helpful for real-time immune surveillance.

X. Wang *et al.* [36] proposed a solution to recognize cancer at earlier stages. It performs pattern classification using the SVM classifier to identify infected cells in the immune system and establishes a relation between tumor genesis and infected cell patterns. Analysis shows that it offers higher detection accuracy for cancer detection and is an appropriate therapy that can be adapted for diagnosis.

EXPERIMENTAL SETUP

For experiments, 100 sample images from kaggle diabetic retina therapy [4] were used with opencv 4.x computer vision python library [37 - 41] over Linux platform, with Intel core i7, 32GB RAM and 1TB space and NVidia graphics card (12GB).

Fig. (7) shows the basic steps to generate a training model for the recognition of multiple patterns in sample data.

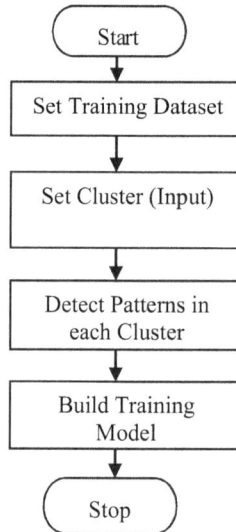

```
          ┌──────────┐
          │  Start   │
          └──────────┘
               │
   ┌───────────────────────┐
   │  Set Training Dataset  │
   └───────────────────────┘
               │
   ┌───────────────────────┐
   │  Set Cluster (Input)   │
   └───────────────────────┘
               │
   ┌───────────────────────┐
   │   Detect Patterns in   │
   │     each Cluster       │
   └───────────────────────┘
               │
   ┌───────────────────────┐
   │    Build Training      │
   │        Model           │
   └───────────────────────┘
               │
          ┌──────────┐
          │   Stop   │
          └──────────┘
```

Fig. (7). Training model preparation.

Step 1: First of all, dataset is prepared and a unique label is assigned to each input image.

Step 2: After that KNN clustering is performed over input images to split an input image into multiple clusters.

Step 3: Finally, in each cluster, patterns are identified using contour and it produces multiple patterns with respect to single input image as shown in Figure 8.

Step 4: After extracting the patterns, a training model is prepared to perform test over other sample images. Following are the extracted patterns from a sample image:

Fig. (**8**) shows the Input retina Image from Kaggle diabetic retina therapy dataset [4].

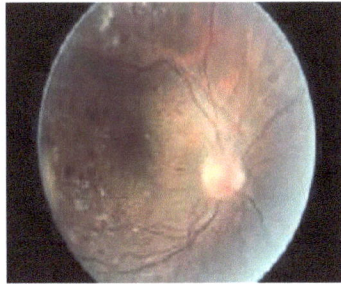

Fig. (8). Input retina image.

Fig. (**9**) shows the different patterns extracted from input image (a diabetic retina image).

Recognized Patterns		Description
Figure 8 (i)		Figure 8 (i) shows the affected region of the diabetic retina.
Figure 8 (ii)		Figure 8 (ii) highlights the effect of diabetes on the retina.
Figure 8(iii)		Figure 8(iii) shows the pattern of extracted retina region w.r.t. Figure 8(i).
Figure 8(iv)		Figure 8(iv) shows multiple retina vessels.

Fig. (9). Recognized patterns from input image.

Fig. (**10**) shows the preparation of testing model.

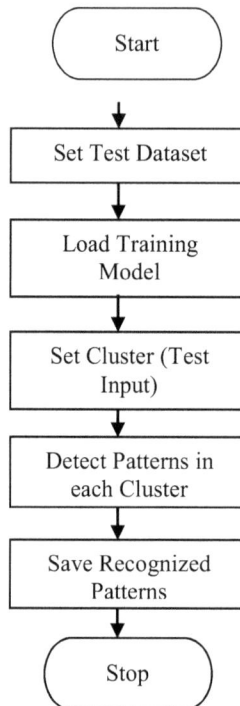

Fig. (10). Testing model preparation.

Step 1: First of all, it initializes the input samples for testing.

Step 2: Training model is also initialized (which was prepared in the 'Training Model preparation' step). It forms clusters and recognizes the patterns withrespect to the existing knowledge base (as described in earlier steps).

Step 3: After above steps, finally, its outcome is saved for further analysis.

Fig. (**11**) shows the comparison of accuracy of two different pattern recognition schemes *i.e.* CBPR and SBPR schemes. As per the results, it can be observed thatthe CBPR scheme has higher accuracy (0.780731959) as compared to the SBPR scheme (0.742903).

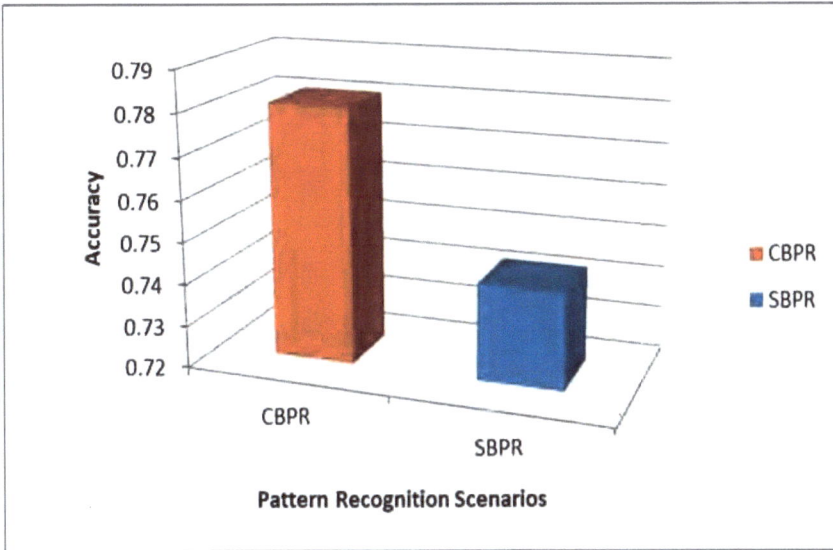

Fig. (11). Comparison-accuracy.

Fig. (**12**) shows the comparison of sensitivity of two different pattern recognition schemes *i.e.* CBPR and SBPR schemes. As per the results, it can be observed that CBPR scheme has higher sensitivity (0.758944612) as compared to the SBPR scheme (0.663176696).

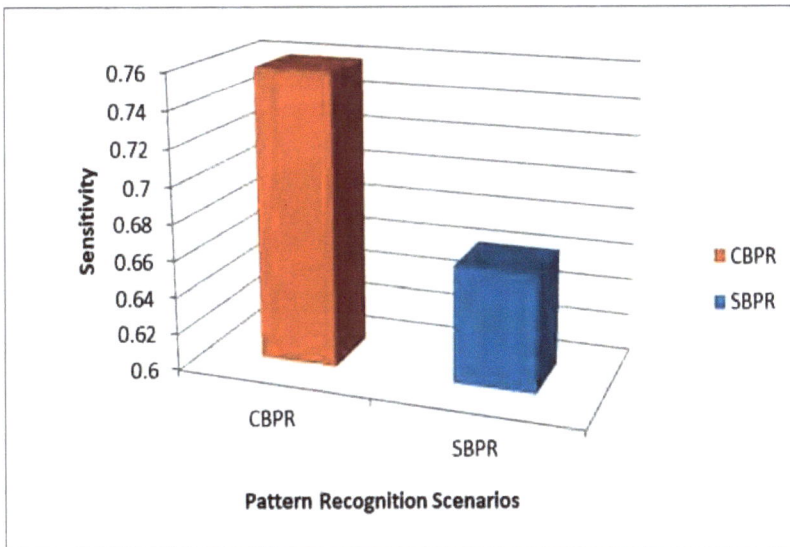

Fig. (12). Comparison-sensitivity.

Fig. (**13**) shows the comparison of specificity of two different pattern recognition schemes *i.e.* CBPR and SBPR schemes. As per the results, it can be observed that it is higher for the CBPR scheme (0.978032) as compared to the SBPR scheme (0.776888).

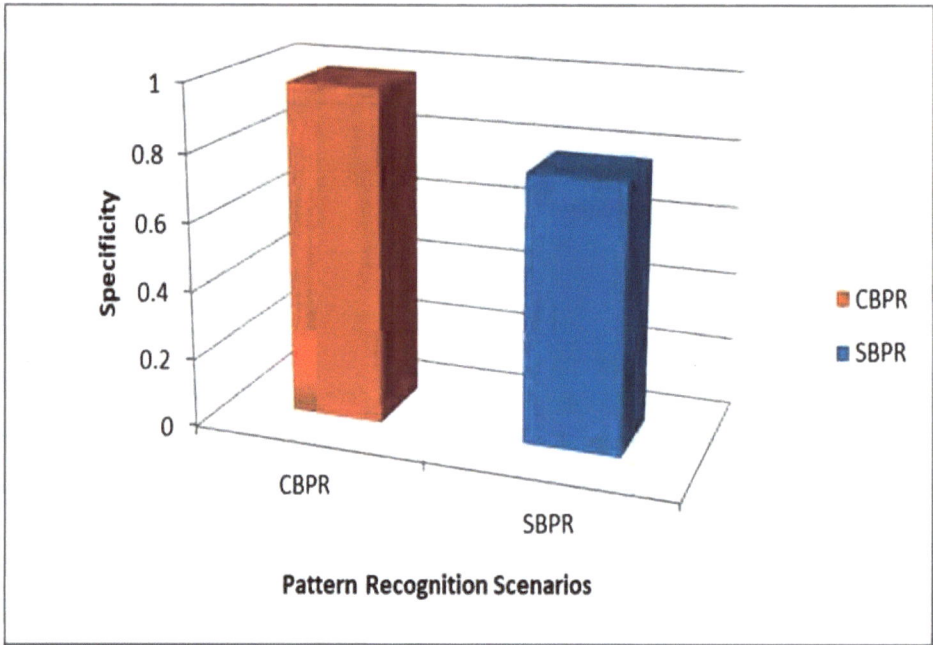

Fig. (13). Comparison-specificity.

CONCLUSION

In this paper, the importance of pattern recognition for medical imagery was investigated and a cluster-based pattern recognition scheme was introduced that builds a machine learning model to create the knowledgebase of retina patterns, called the training model, and later on it is used for testing the different retina patterns with respect to diabetic retina samples.

Its performance was also measured using different parameters (pattern recognition accuracy/sensitivity/specificity).

Outcomes show that CBPR delivered higher pattern recognition accuracy, its sensitivity is also better than SBPR, and the specificity of CBPR is marginally higher than SBPR.

Currently, it is developed to recognize patterns of retina samples only and it will be further enhanced to recognize patterns of other medical images (ECG/MRI, *etc*.) also.

REFERENCES

[1]   D. Kang, S. Kim, Y. Jung, and H.S. Ryoo, "Generating interpretable patterns for biomedical image classification", *IEEE International Conference on Bioinformatics and Biomedicine (BIBM), IEEE,* 2021pp. 1658-1660
[http://dx.doi.org/10.1109/BIBM52615.2021.9669323]

[2]   Available from: https://wiki.cancerimagingarchive.net

[3]   Available from: https://stanfordmlgroup.github.io

[4]   Available from: https://www.kaggle.com

[5]   Available from: https://ecglibrary.com

[6]   S. F. Ershad, and S. Ramakrishnan, "Cervical cancer diagnosis based on modified uniform local ternary patterns and feed forward multilayer network optimized by genetic algorithm", *Computers in Biology and Medicine,* vol. 144, p. 105392, 2022.

[7]   Z. Li, J. Yang, Y. Wang, M. Cai, X. Liu, and K. Lu, "Early diagnosis of Parkinson's disease using Continuous Convolution Network: Handwriting recognition based on off-line hand drawing without template", *J. Biomed. Inform.,* vol. 130, p. 104085, 2022.
[http://dx.doi.org/10.1016/j.jbi.2022.104085] [PMID: 35490964]

[8]   A.P.B. Budiarsa, J.S. Leu, K.K.F. Yuen, and X. Sigalingging, "Improved swarm-wavelet based extreme learning machine for myoelectric pattern recognition", *Biomed. Signal Process. Control,* vol. 77, p. 103737, 2022.
[http://dx.doi.org/10.1016/j.bspc.2022.103737]

[9]   H. Zhang, X. Wang, Y. Zhang, G. Cao, and C. Xia, "Design on a wireless mechanomyography acquisition equipment and feature selection for lower limb motion recognition", *Biomedical Signal Processing and Control,* vol. 77, p. 103679, 2022.
[http://dx.doi.org/10.1016/j.bspc.2022.103679]

[10]   A. Subasi, S. S. Panigrahi, B. S. Patil, M. A. Canbaz, and R. Klén, "Advanced pattern recognition tools for disease diagnosis", *Intelligent Data-Centric Systems, 5G IoT and Edge Computing for Smart Healthcare, Academic Press,* p. 195-229, 2022.
[http://dx.doi.org/10.1016/B978-0-323-90548-0.00011-5]

[11]   D.M. Radha Devi, P.L. Jancy, P. Tamilselvi, and V. Aishwarya, "Patient History Tracking using Local Binary Pattern Histogram(LBPH) Algorithm", *International Conference on Communication, Computing and Internet of Things (IC3IoT),* p. 1-6, 2022.
[http://dx.doi.org/10.1109/IC3IOT53935.2022.9767862]

[12]   J.A. Patel, and M. Patel, "Different Transfer Learning Approaches for Recognition of Lung Sounds: Review", *2nd International Conference on Artificial Intelligence and Smart Energy (ICAIS),* pp. 738-742, 2022.
[http://dx.doi.org/10.1109/ICAIS53314.2022.9742754]

[13]   T. Tuncer, S. Dogan, P. Plawiak, and A. Subasi, "A novel Discrete Wavelet-Concatenated Mesh Tree and ternary chess pattern based ECG signal recognition method", *Biomedical Signal Processing and Control,* vol. 72, p. 1-8, 2022.
[http://dx.doi.org/10.1016/j.bspc.2021.103331]

[14]   S. Qiao, Q. Yu, Z. Zhao, L. Song, H. Tao, T. Zhang, and C. Zhao, "Edge extraction method for medical images based on improved local binary pattern combined with edge-aware filtering", *Biomedical Signal Processing and Control,* vol. 74, p. 103490, 2022.
[http://dx.doi.org/10.1016/j.bspc.2022.103490]

[15]   U. Ogiela, and V. Snášel, "Predictive intelligence in evaluation of visual perception thresholds for visual pattern recognition and understanding", *Information Processing & Management,* vol. 59, no. 2, p. 102865, 2022.
[http://dx.doi.org/10.1016/j.ipm.2022.102865]

[16] K. Yuki, and S. Koutsogiannaki, "Pattern recognition receptors as therapeutic targets for bacterial, viral and fungal sepsis", *International Immunopharmacology,* vol. 98, pp. 1-7, 2021.
[http://dx.doi.org/10.1016/j.intimp.2021.107909]

[17] A. I. Korda, C. Andreou, and S. Borgwardt, "Pattern classification as decision support tool in antipsychotic treatment algorithms", *Experimental Neurology,* vol. 339, pp. 1-8, 2021.
[http://dx.doi.org/10.1016/j.expneurol.2021.113635]

[18] K. Kitagawa, T. Nagasaki, S. Nakano, M. Hida, S. Okamatsu, and C. Wada, "Comparison of machine learning algorithms for patient handling recognition based on body mechanics", *3rd Global Conference on Life Sciences and Technologies (LifeTech),* pp. 77-79, 2021.
[http://dx.doi.org/10.1109/LifeTech52111.2021.9391968]

[19] H. Gunasinghe, J. McKelvie, A. Koay, and M. Mayo, "Comparison of pretrained feature extractors for glaucoma detection", *18th International Symposium on Biomedical Imaging (ISBI),* pp. 390-394, 2021.
[http://dx.doi.org/10.1109/ISBI48211.2021.9434082]

[20] O. Leombruni, A. Annovi, P. Giannetti, N. V. Biesuz, C. Roda, and M. Cal, "Pattern-matching unit for medical applications", *IEEE Transactions on Nuclear Science,* vol. 68, no. 8, pp. 2140-2145, 2021.
[http://dx.doi.org/10.1109/TNS.2021.3083894]

[21] R. Ahsan, M. R. Tahsili, F. Ebrahimi, E. Ebrahimie, and M. Ebrahimi, "Image processing unravels the evolutionary pattern of SARS-CoV-2 against SARS and MERS through position-based pattern recognition", *Computers in Biology and Medicine,* vol. 134, pp. 1-11, 2021.
[http://dx.doi.org/10.1016/j.compbiomed.2021.104471]

[22] D. Gómez-Andrés, A. Oulhissane, and S. Quijano-Roy, "Two decades of advances in muscle imaging in children: from pattern recognition of muscle diseases to quantification and machine learning approaches", *Neuromuscul. Disord.,* vol. 31, no. 10, pp. 1038-1050, 2021.
[http://dx.doi.org/10.1016/j.nmd.2021.08.006] [PMID: 34736625]

[23] Y. Ma, Z. Li, J. Gou, L. Ding, D. Yang, and G. Feng, "Adoption of improved neural network blade pattern recognition in prevention and control of corona virus disease-19 pandemic", *Pattern Recognition Letters,* vol. 151, pp. 275-280, 2021.
[http://dx.doi.org/10.1016/j.patrec.2021.08.033]

[24] L. Borne, D. Rivière, A. Cachia, P. Roca, C. Mellerio, C. Oppenheim, and J. F. Mangin, "Automatic recognition of specific local cortical folding patterns", *NeuroImage,* vol. 238, pp. 1-12, 2021.

[25] F. Loncaric, P. M. M. Castellote, S. S. Martinez, D. Fabijanovic, L. Nunno, M. Mimbrero, L. Sanchis, A. Doltra, S. Montserrat, M. Cikes, F. Crispi, G. Piella, M. Sitges, and B. Bijnens, "Automated pattern recognition in whole-cardiac cycle echocardiographic data: capturing functional phenotypes with machine learning", *Journal of the American Society of Echocardiography,* vol. 34, no. 11, pp. 1170-1183, 2021.
[http://dx.doi.org/10.1016/j.echo.2021.06.014]

[26] C. Djellali, "A data-driven deep learning model to pattern recognition for medical diagnosis, by using model aggregation and model selection", *Procedia Computer Science,* vol. 177, pp. 387-395, 2020.

[27] S.A.V. Begum, and M.P. Rani, "Recognition of neurodegenerative diseases with gait patterns using double feature extraction methods", *4th International Conference on Intelligent Computing and Control Systems (ICICCS),* pp. 332-338, 2020.

[28] I. Urbaniak, and M. Wolter, "Quality assessment of compressed and resized medical images based on pattern recognition using a convolutional neural network", *Communications in Nonlinear Science and Numerical Simulation,* vol. 95, pp. 1-29, 2021.
[http://dx.doi.org/10.1016/j.cnsns.2020.105582]

[29] M. Liu, M. Zhou, T. Zhang, and N. Xiong, "Semi-supervised learning quantization algorithm with deep features for motor imagery EEG Recognition in smart healthcare application", *Applied Soft Computing,* vol. 89, p. 106071, 2020.

[30] Y. Jiang, C. Chen, X. Zhang, C. Chen, Y. Zhou, G. Ni, S. Muh, and S. Lemos, "Shoulder muscle activation pattern recognition based on sEMG and machine learning algorithms, Computer Methods and Programs", *Biomedicine,* vol. 197, p. 105721, 2020.

[31] P. Feng, and L. Feng, "Sequence based prediction of pattern recognition receptors by using feature selection technique", *International Journal of Biological Macromolecules,* vol. 162, pp. 931-934, 2020.
[http://dx.doi.org/10.1016/j.ijbiomac.2020.06.234]

[32] Y. J. C-Pino, M. C. González, V. Quintana-Peña, J. Valderrama, B.Muñoz, J. Orozco, A. Navarro, "Automatic Gait Phases Detection in Parkinson Disease: A Comparative Study", *42nd Annual International Conference of the IEEE Engineering in Medicine & Biology Society,* pp. 798-802, 2020.

[33] K. Vatanparvar, E. Nemati, V. Nathan, M.M. Rahman, and J. Kuang, "CoughMatch – Subject Verification Using Cough for Personal Passive Health Monitoring", *42nd Annual International Conference of the IEEE Engineering in Medicine & Biology Society (EMBC),* pp. 5689-5695, .
[http://dx.doi.org/10.1109/EMBC44109.2020.9176835]

[34] J. Waldthaler, C. K. Zechlin, L. Stock, Z. Deeb, and L. Timmermann, "New insights into facial emotion recognition in Parkinson's disease with and without mild cognitive impairment from visual scanning patterns", *Clinical Parkinsonism & Related Disorders,* vol. 1, pp. 102-108, 2019.
[http://dx.doi.org/10.1016/j.prdoa.2019.11.003]

[35] L. Bai, W. Li, W. Zheng, D. Xu, N. Chen, and J. Cui, "Promising targets based on pattern recognition receptors for cancer immunotherapy", *Pharmacol. Res.,* vol. 159, pp. 1-13, 2020.
[http://dx.doi.org/10.1016/j.phrs.2020.105017]

[36] X. Wang, W. Shang, X. Li, and Y. Chang, "Methylation signature genes identification of cancers occurrence and pattern recognition", *Comp. Biol. Chem.,* vol. 85, pp. 1-7, 2020.
[http://dx.doi.org/10.1016/j.compbiolchem.2019.107198]

[37] M. Hu, H. Lin, Z. Fan, W. Gao, L. Yang, C. Liu, and Q. Song, "Learning to recognize chest xray images faster and more efficiently based on multi-kernel depthwise convolution", *IEEE Access,* vol. 8, pp. 37265-37274, 2020.
[http://dx.doi.org/10.1109/ACCESS.2020.2974242]

[38] E. Hoppe, J. Wetz, S.S. Yoon, M. Bacher, P. Roser, B. Stimpe, A. Preuhs, and A. Maier, "Deep learning-based ECG-free cardiac navigation for multi-dimensional and motion-resolved continuous magnetic resonance imaging", *IEEE Transactions on Medical Imaging,* vol. 40, no. 8, pp. 2105-2117, 2021.
[http://dx.doi.org/10.1109/TMI.2021.3073091]

[39] M. Sabouri, G. Hajianfar, M. Amini, Z. Hosseini, S. Madadi, T. Ghaedian, M. Ghassed, F. Rastgou, and A.B. Rajabi, "I, Shiri, H. Zaidi, "Cardiac Pattern Recognition from SPECT Images Using Machine Learning Algorithms", *IEEE Nuclear Science Symposium and Medical Imaging Conference (NSS/MIC), IEEE,* pp. 1-3, 2021.

[40] Available from: https://opencv.org/

[41] A. Chopra, D. C. Verma, R. Gujral, Machine Learning-Based Active Contour Approach for The Recognition of Brain Tumor Progression, Book Chapter: Data Science for Effective Healthcare Systems, Data Science for Effective Healthcare Systems, 1st Edition, Routledge, CRC Press, Taylor & Francis, pp. 183-198, 2022.

CHAPTER 7

AutoMate: Ubiquitous Smart Home System using Arduino and ESP8266 Module

Rakhi Kamra[1,*] and **Soumya Chaudhary**[1]

[1] *Department of Electrical and Electronics Engineering, Maharaja Surajmal Institute of Technology, Delhi, India*

Abstract: This research paper proposes a versatile standalone, cost-effective smart home system that does not require any substantial changes to the existing framework. The project is built with Arduino Uno and NodeMCU (ESP8266) microcontrollers that operate two distinct 4-channel Relays, which in turn control household appliances. Ubiquitous computing, also known as pervasive computing, is a computer science term that refers to the ability to be present everywhere and at any time. According to this notion, a user may interact with computers, which may exist in many forms such as laptops, tablets, and terminals in everyday items. To demonstrate the feasibility and efficacy of the proposed smart home system, devices such as LED lights, power connectors, and a fan have been integrated into the system. The NodeMCU is programmed using the Arduino IDE. It is linked to the Internet, where it receives signals and carries out the user-programmed actions on the relay. By clicking a button on the mobile application's interface, this function enables users to manually control all of their home appliances.

Keywords: Home automation, Internet of things, IoT, Pervasive, Smart home, Ubiquitous computing.

INTRODUCTION

Nearly 15 years ago, Mark Weiser outlined his vision of ubiquitous computing, which included computer systems that would blend into the background of our daily lives and computing infrastructure that would be readily available to everyone. Weiser predicts that as computing technology advances, it will become as normal to use in daily life as, for example, writing on paper with a pencil. By integrating gadgets and appliances into the environment, these improvements will create new settings that are replete with computer and communication capabilities while disguising them from the user.

* **Corresponding author Rakhi Kamra:** Department of Electrical and Electronics Engineering, Maharaja Surajmal Institute of Technology, Delhi, India; E-mail: rakhikamra@msit.in

Neha Kishore, Pankaj Nanglia, Shilpa Gupta & Ashutosh Kumar Dubey (Eds.)

The expanding use of computing technology in numerous sectors of life creates new opportunities as well as challenges for computer scientists in various disciplines. We can create smart environments that benefit the user in a variety of ways by using connected devices. The miniaturisation of sensors and actuators incorporated into prevalent devices allows for a progressive paradigm shift towards "ubiquitous computing".

Our personal living environment, in particular, is rapidly becoming the center of our attention. Flats and houses are being transformed into so-called smart homes. The purpose of these smart homes is to help residents achieve greater comfort, safety, and energy efficiency. This approach is supported by several technologies; nevertheless, there is a lack of a unified framework that addresses these issues. The main obstacles are the integration of multiple devices using various communication protocols, the acquisition of information from the environment and its aggregated delivery, and the development of intuitively useful services [1 - 7].

The conceptualization of a smart home system must take into account a number of factors. The system must be user-friendly, scalable, and affordable so that additional devices can be quickly integrated into it. This paper proposes a low-cost wireless controlled smart home system, AutoMate, for managing as well as monitoring the indoor environment. An Android-based app, accessible from any device that supports Android, is used to remotely access and operate appliances and other devices using an embedded micro-web server with an authentic IP connection. The micro web server on the Arduino Ethernet replaces the PC, and the system requires authentication from the user to access it.

RELATED LITERATURE

The concept of a "smart home" is not new to science society, but it is still quite distant from the public's perception and expectation. Home automation is a growing topic as electronic technologies converge. There have been a number of smart systems developed where the control is by Bluetooth [8 - 13], the Internet [14 - 16], SMS-based [17], *etc.* The majority of modern laptops, tablets, and mobile phones include built-in adapters, which improve Bluetooth capabilities while also indirectly lowering system costs. However, it restricts the control to the Bluetooth-enabled environment, whereas the majority of other technologies are not practical to deploy as low-cost solutions.

A Wi-Fi-based home automation system is described [18]. It controls the connected home gadgets *via* a PC-based web server with a built-in Wi-Fi card. Users have the option of controlling and managing the system locally (LAN) or remotely (internet). The system is compatible with a wide range of home automation devices, including security and power management components. A similar architecture is put out in [19], with the home agent operating on a PC coordinating the tasks.

In other articles [20, 21, 23], internet-controlled systems with a dedicated web server, a database, and a web page for connecting and controlling the devices were also shown. These systems need a PC, which directly raises the price and consumption of energy. However, there will be additional charges associated with the creation and hosting of the website.

A study [22] presents the design and construction of a voice-activated wireless automation system that uses a microcontroller. Through a microphone, the user gives voice instructions, which are then processed and wirelessly transmitted *via* a radio frequency (RF) connection to the main control receiver unit. The characteristics of the speech command are extracted using a voice recognition module. The microcontroller then processes this extracted signal to carry out the intended operation. The limitation that the system can only be managed from inside the RF range is a disadvantage. Research works in [24 - 26] have designed their wireless-based home automation system, which is a less complex, affordable smart home system that will control many actuators from the installed sensors data. Once more, a PC is utilised, which results in higher costs and consumption of energy.

SYSTEM DESIGN

This section presents the proposed smart home system's sophisticated layout as well as its component parts.

System Architecture

Fig. (**1**) depicts a brief overview of the developed system's architecture. The system comprises an Arduino Ethernet-based micro web-server and a Wi-Fi module ESP8266, along with a USB to TTL converter, that works on the AT command set for communication. The Arduino microcontroller is the primary controller that hosts the micro web server and executes the essential tasks. The actuators/relays are linked directly to the main controller. By using TCP protocols to interface with telnet, which engages with the tiny web server *via* a connection, the smart home ecosystem may be managed and monitored remotely. On the user's device, any internet connection through Wi-Fi or a 3G/4G network can be utilised.

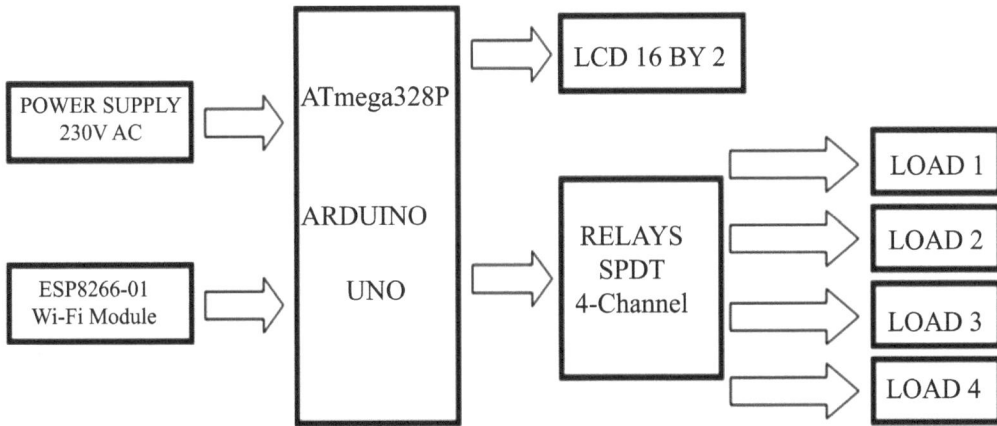

Fig. (1). System architecture of the proposed automate: ubiquitous smart home.

The features offered by the suggested design include the ability to control energy management systems like LED lighting, fans, and two additional power outlets remotely *via* Wi-Fi and an Android app. The digital input/output pins of the Arduino Uno are used to connect these devices. The ESP8266-01 communication module is used to link these devices to the local Wi-Fi network, allowing the If Then, Then That (IFTTT) service to be potentially implemented, and it provides user authentication to access the smart home system using the Telnet application.

Software Development

From the official Arduino website, first, we have to download the Arduino software IDE. Install the downloaded programme on your computer. The code for AutoMate will be written in Fig. (**2**) and uploaded to the ATmega 328 controller.

RESULTS AND DISCUSSION

The suggested smart home system in this paper underwent thorough development and testing in order to demonstrate that it is both practical and efficient. The solution offers a cost-effective hardware prototype that aids in saving power, making it an energy-efficient system. The TELNET application helps the user gain insights that can be used to make further modifications. The screenshots of the Telnet application from a mobile device are shown in Fig. (**3**). Write the IP address and port in the programme to connect to the Wi-Fi, as illustrated in Figs. (**4** to **6**) below. Therefore, a user can effortlessly control the lights in any room using a specific keyword. The smart home system is accessible only after authentication, as previously indicated. If the proper authentication is given, the app then displays the smart home controls page and displays a message indicating login success.

```
     AutoMate.ino              ReadMe.adoc        ▼
1   #include <SoftwareSerial.h>  //Including the software serial library
2   #define DEBUG true
3   SoftwareSerial esp8266(2,3); // This will make the Arduino pin 2 as the RX pin and pin 3 as the TX. Software UART
4 ▾ /* So you have to connect the TX of the esp8266 to the pin 2 of the Arduino and the TX of the esp8266 to the
5   pin 3 of the Arduino. This means that you need to connect the TX line from the esp to the Arduino's pin 2 */
6
7   void setup()
8 ▾ {
9     Serial.begin(9600);     // Setting the baudrate to 9600
10    esp8266.begin(9600);    // Set it according to your esp's baudrate. Different esp's have different baud rates.
11    pinMode(11,OUTPUT);     // Setting the pin 11 as the output pin.
12    digitalWrite(11,LOW);   // Making it low.
13    pinMode(12,OUTPUT);     // Setting the pin 12 as the output pin..
14    digitalWrite(12,LOW);   // Making pin 12 low.
15    pinMode(13,OUTPUT);     // Setting the pin 13 as the output pin.
16    digitalWrite(13,LOW);   // Making pin 13 low.
17    sendData("AT+RST\r\n",2000,DEBUG);          //This command will reset module to default
18    sendData("AT+CWMODE=2\r\n",1000,DEBUG);     // This will configure the mode as access point
19    sendData("AT+CIFSR\r\n",1000,DEBUG);        // This will get ip address and will show it
20    sendData("AT+CIPMUX=1\r\n",1000,DEBUG);     // This will configure the ESP8266 for multiple connections
21    sendData("AT+CIPSERVER=1,80\r\n",1000,DEBUG); // This will set the server on port 80
22  }
23  void loop()
24 ▾ {
25    if(esp8266.available()) // Checking that whether the esp8266 is sending a message or not (Software UART Data)
26 ▾  {
27      if(esp8266.find("+IPD,"))
28 ▾    {
29        delay(1000);         // Waiting for 1 sec
```

```
     AutoMate.ino              ReadMe.adoc        ▼
30        int connectionId = esp8266.read()-48;   // Subtracting 48 from the character to get the number.
31        esp8266.find("pin=");                    // Advancing the cursor to the "pin "
32        int pinNumber = (esp8266.read()-48)*10;  // Getting the first number which is pin 13
33        pinNumber +  (esp8266.read()-48);        // This will get the second number. For example, if pin number is 13
34                                                  //then the 2nd number will be 3 and then add it to the first number
35        digitalWrite(pinNumber, !digitalRead(pinNumber)); // This will toggle the pin
36        // The following commands will close the connection
37        string closeCommand = "AT+CIPCLOSE=";
38        closeCommand+ connectionId;
39        closeCommand+="\r\n";
40        sendData(closeCommand,1000,DEBUG);       // Sending the data to the ESP8266 to close the command
41      }
42    }
43  }
44  String sendData(String command, const int timeout, boolean debug) // function to send the data to the esp8266
45 ▾ {
46    string response = "";
47    esp8266.print(command);              // Send the command to the ESP8266
48    long int time = millis();
49    while( (time+timeout) > millis()) // ESP8266 will wait for some time for the data to receive
50 ▾  {
51      while(esp8266.available())         // Checking whether ESP8266 has received the data or not
52      { char c = esp8266.read();         // Read the next character.
53        response+=c;                      // Storing the response from the ESP8266
54      }
55    }
56    if(debug)
57      Serial.print(response);            // Printing the response of the ESP8266 on the serial monitor.
58    return response;
59  }
```

Fig. (2). Screenshots of code written in arduino ide to program ATmega 328P.

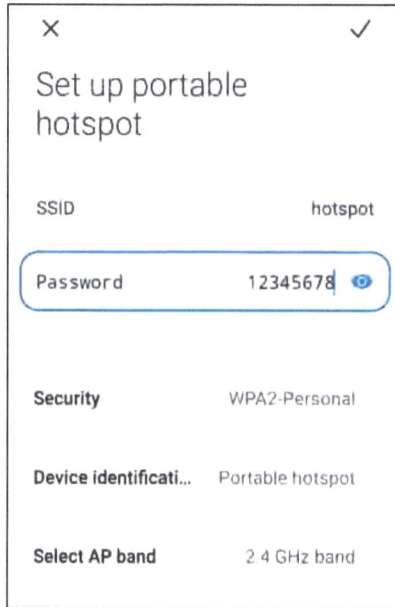

Fig. (3). Screenshot of the portable hotspot being set up.

Fig. (4). Screenshot of the IP address and port for connecting purposes on telnet.

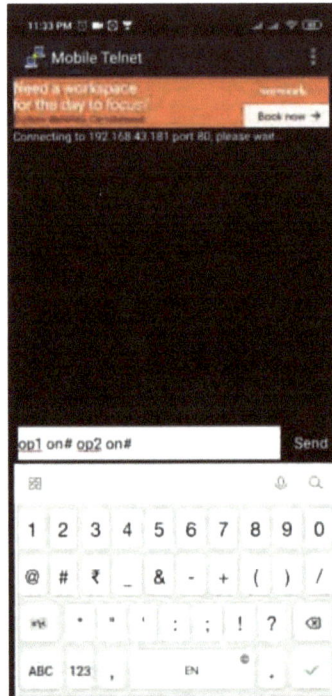

Fig. (5). Screenshot of the commands to switch on the devices on telnet.

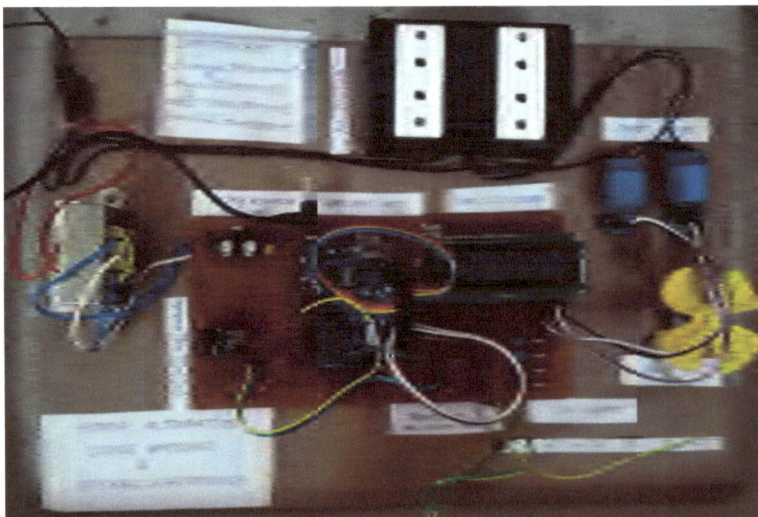

Fig. (6). Working hardware prototype of automate smart home system.

CONCLUSION

The paper proposes and implements an internet-based smart home system that is operated remotely after user verification. The internet is used by the Android-based smart home software, Telnet, to interact with the web server. Arduino is used for the programming task due to the platform being open-source and having great user support. The distance limitation is overcome by using the ESP8266 Wi-Fi module for better range instead of a Bluetooth module. Using this module ensures great scalability and flexibility of the proposed system, along with prompt notification delivery. Any Android-compatible smartphone that can run the software may be used to control and monitor the smart home environment. A cost-efficient system has been created that does not need a PC because the microcontroller handles all processing and controlling all the devices from one place makes the system highly convenient for the user. Future work may include enabling SMS and call alerts, minimising the need for wiring modifications when installing the suggested system in existing homes, and developing a wireless network for managing and monitoring the smart home environment.

ACKNOWLEDGEMENT

Financial support for this project was provided by the Department of Electrical and Electronics Engineering, Maharaja Surajmal Institute of Technology, Delhi, India. The authors wish to thank the department for the kind and generous support, which helped in the purchasing of the materials for conducting this research.

REFERENCES

[1]   M. Tahir, "Wi-Fi Aided Home Energy Management System and AC Prediction through Temperature and Humidity Sensors", *2022 International Conference on Cyber Resilience (ICCR)*, 2022.
[http://dx.doi.org/10.1109/ICCR56254.2022.9995822]

[2]   Orfanos, Vasilios A., *et al.* "A comprehensive review of iot networking technologies for smart home automation applications." *Journal of Sensor and Actuator Networks* 12.2 (2023): 30. D. Sarunyagate, Ed., Lasers. NewYork: McGraw-Hill, 1996.
[http://dx.doi.org/10.3390/jsan12020030]

[3]   Ramkumar, A., *et al.* "Android Controlled Smart Home Automation with Security System." 2022 2nd International Conference on Advance Computing and Innovative Technologies in Engineering (ICACITE). IEEE, 2022. K. Kimura and A. Lipeles, "Fuzzy controller component", U.S. Patent 14, 860, 040, December 14, 1996.

[4]   Sheshalani Balasingam, M.K. Zapiee, and D. Mohana, "Smart Home Automation System Using IOT", *International Journal of Recent Technology and Applied Science,* vol. 4, no. 1, pp. 44-53, 2022.
[http://dx.doi.org/10.36079/lamintang.ijortas-0401.332]

[5]   B. Mallikarjuna, "Feedback-based resource utilization for smart home automation in fog assistance IoT-based cloud", *Research Anthology on Cross-Disciplinary Designs and Applications of Automation. IGI Global,* pp. 803-824, 2022.
[http://dx.doi.org/10.4018/978-1-6684-3694-3.ch039]

[6]   M.S.H. Shawon, "Voice Controlled Smart Home Automation System Using Bluetooth Technology", *2021 4th International Conference on Recent Trends in Computer Science and Technology (ICRTCST),* 2022.

[7]   N. Alsbou, N.M. Thirunilath, and I. Ali, "Smart Home Automation IoT System for Disabled and

Elderly", *Electronics and Mechatronics Conference (IEMTRONICS)*, 2022.
[http://dx.doi.org/10.1109/IEMTRONICS55184.2022.9795738]

[8]   S. Anwaarullah, and S. V. Altaf, "RTOS based Home Automation System using Android", *International Journal of Advanced Trends in Computer Science and Engineering*, vol. 2, pp. 480-484, 2013.

[9]   C. Chiu-Chiao, H. C. Yuan, W. Shiau-Chin, and L. Cheng-Min, "Bluetooth-Based AndroidInteractive Applications for Smart Living", *2nd International Conferenceon Innovations in Bioinspired Computing and Appplications (IBICA 2011)*, pp. 309-312, 2011.

[10]   D. Javale, M. Mohsin, S. Nandanwar, and M. Shingate, "Home Automation and Security SystemUsing Android ADK", *International Journal of Electronics Communication and ComputerTechnology (IJECCT)*, vol. 3, pp. 382-385, 2013.

[11]   J. Potts, and S. Sukittanon, "Exploiting Bluetooth on Android mobile devices for home securityapplications", *Southeastcon, 2012 Proceedings of IEEE Orlando, FL*, 2012.

[12]   R. A. Ramlee, M. H. Leong, R. S. S. Singh, M. M. Ismail, M. A. Othman, and H. A. Sulaiman, "Bluetooth remote home automation system using android application", *The International Journal of Engineering and Science*, vol. 2, pp. 149-153, 2013.

[13]   M. Yan, and H. Shi, "Smart Living Using Bluetooth Based Android Smartphone", *International Journal of Wireless & Mobile Networks*, vol. 5, pp. 65-72, 2013.
[http://dx.doi.org/10.5121/ijwmn.2013.5105]

[14]   B.M. Chen, Shaoyan Hu, V. Ramakrishnan, Yuan Zhuang, Jianping Chen, and Y. Zhuang, "A web-based virtual laboratory on a frequency modulation experiment", *IEEE Trans. Syst. Man Cybern. C*, vol. 31, no. 3, pp. 295-303, 2001.
[http://dx.doi.org/10.1109/5326.971657]

[15]   N. Swamy, O. Kuljaca, and F.L. Lewis, "Internet-based educational control systems lab using NetMeeting", *IEEE Trans. Educ.*, vol. 45, no. 2, pp. 145-151, 2002.
[http://dx.doi.org/10.1109/TE.2002.1013879]

[16]   K.K. Tan, T.H. Lee, and C.Y. Soh, "Internet-based monitoring of distributed control systems-An undergraduate experiment", *IEEE Trans. Educ.*, vol. 45, no. 2, pp. 128-134, 2002.
[http://dx.doi.org/10.1109/TE.2002.1013876]

[17]   M.S.H. Khiyal, A. Khan, and E. Shehzadi, "SMS Based Wireless Home Appliance ControlSystem (HACS) for Automating Appliances and Security", *Issues in Informing Science and Information Technology*, vol. 6, pp. 887-894, 2009.

[18]   A. ElShafee, and K.A. Hamed, "Design and Implementation of a WiFi Based Home AutomationSystem", *World Acad. Sci. Eng. Technol.*, vol. 68, pp. 2177-2180, 2012.

[19]   R.D. Caytiles, and B. Park, "Mobile IP-Based Architecture for Smart Homes", *Int. J. Smart Home*, vol. 6, pp. 29-36, 2012.

[20]   A.Z. Alkar, and U. Buhur, "An internet based wireless home automation system for multifunctional devices", *IEEE Trans. Consum. Electron.*, vol. 51, no. 4, pp. 1169-1174, 2005.
[http://dx.doi.org/10.1109/TCE.2005.1561840]

[21]   N-S. Liang, L-C. Fu, and C-L. Wu, "An integrated, flexible, and Internet-based controlarchitecture for home automation system in the Internet era", *IEEE International Conference onRobotics and Automation*, pp. 1101-1106, 2002.Washington, DC

[22]   A. Rajabzadeh, A.R. Manashty, and Z.F. Jahromi, "A Mobile Application for Smart HouseRemote Control System", *World Academy of Science, Engineering and Technology*, vol. 62, 2010.

[23]   T. Manish, B. Nagaraju, R. Kamra, R.C. Tanguturi, and D.N. Sahu, "An Intelligent Internet of Medical Things based Smart Healthcare Monitoring Equipment for Patients", *European Chemical Bulletin*, vol. 12, no. 4, pp. 1048-1065, 2023.

[24]   U. Sharma, and S.R.N. Reddy, "Design of Home/Office Automation Using Wireless SensorNetwork", *Int. J. Comput. Appl.,* vol. 43, pp. 53-60, 2012.

[25]   K.P. Dutta, P. Rai, and V. Shekher, "Microcontroller Based Voice Activated Wireless AutomationSystem", *VSRD Internation Journal of Electrocal, Electronics & Communication Engineering,* vol. 2, pp. 642-649, 2012.

[26]   S.S. Tippannavar, "N. Shivaprasad, and Praveen Kumar. "Smart Home Automation Implemented using LabVIEW and Arduino", *2022 International Conference on Electronics and Renewable Systems (ICEARS),* 2022.
[http://dx.doi.org/10.1109/ICEARS53579.2022.9752265]

Digital Forensics in Mobile Phones: An Overview of Data Acquisition Techniques and its Challenges

Neha Kishore[1,*] and **Priya Raina**[2]

[1] *Department of Computer Science and Engineering, Maharaja Agrasen University, Himachal Pradesh, India*

[2] *School of Engineering and Technology, Chitkara University, Himachal Pradesh, India*

Abstract: Over the past decade, advances in hardware, software, and networking have led to the evolution of modern-day smart devices, which are no longer simply mobile phones, but have significant computing power. Such a phenomenal increase in the performance and capabilities of smartphones, tablets, and personal digital assistants, along with the convenience of using them, has practically led them to replace computers and notebooks. However, their small size makes them susceptible to theft. Also, the data they contain coupled with continuous network connectivity makes them susceptible to malicious activities and attacks. Investigation of such incidents as well as the increasing technical difficulties in extracting evidence from mobile devices has resulted in the emergence of mobile forensics within the digital forensics discipline. Mobile forensics is specialized in retrieving and processing evidence from mobile devices such that it is admissible in a court of law. While the scope of mobile forensics includes advanced evidence analysis and threat intelligence to thwart attacks or malicious activities, data acquisition still remains its main focus. This paper presents an overview of the research conducted in the domain of forensic acquisition of mobile phones during the past decade, identifying the challenges and opportunities in the field.

Keywords: Acquisition, Android, Digital forensics, Digital investigations, Evidence, Framework, Mobile forensics, Preservation, Tools.

INTRODUCTION

Over the past decade, mobile devices have evolved in their computing capabilities, thus becoming increasingly ubiquitous and pervasive. These compact yet powerful devices are often associated with the cloud, through various "apps". They are convenient to use and provide users with seamless connectivity, fulfilling our daily computing needs, and literally bringing the world to our fingertips. Our continuous interaction with mobile devices makes them a store-

* **Corresponding author Neha Kishore:** Department of Computer Science and Engineering, Maharaja Agrasen University, Himachal Pradesh, India; E-mail: nehakishore.garg@gmail.com

house of sensitive personal data and associated metadata. On one hand, this makes them vulnerable to attacks by cyber-criminals, who adopt various means, like phishing, social engineering, *etc.*, to con the users, in which case the device would carry the footprints of the attack. On the other hand, these devices can also provide incriminating evidence during criminal investigations, as their ubiquity often causes them to be involved in the crime, directly or indirectly. Therefore, forensic science particularly digital forensics as a field is hugely benefited by the growth of mobile devices.

However, although functionally similar, mobile devices are very different from traditional computers *e.g.*, in hardware, software, storage, mobility, connectivity, power consumption, *etc.* Consequently, mobile devices, like smartphones, PDAs, smart watches, *etc.* are required to be handled differently, using a specialized branch called mobile forensics [1, 2]. Mobile forensics investigation often spans multiple layers.

This chapter presents an overview of the state-of-the-art mobile forensic techniques and discusses challenges and research opportunities, while covering the following aspects:

- Brief description of mobile computing, digital forensics, and the need for mobile forensics.
- Mobile forensics: Discussion on frameworks, tools and recent research.
- Challenges and opportunities in the field of mobile forensics.

MOBILE COMPUTING

Mobile computing refers to the technologies that enable people to communicate and fulfill basic computing needs without a fixed-point connection or location-related restrictions. It includes both, the mobile device (hardware and software) as well as network connectivity. Mobile devices have evolved over the past decade into hand-held computers, thanks to advancements in hardware (touch screen, processor power, memory, battery life). The credit of this evolution also goes to development and standardization of operating systems, primarily Android, iOS, Tizen, *etc.* As of today, 99% of the market is captured by Android and iOS, with Android phones comprising more than 70% of the market share [3 - 5].

Android Platform

The popularity of Android is because of its open-source nature, availability of a wide variety of applications, and its compatibility with a diverse set of hardware, made possible due to its underlying Linux kernel. This allows the manufacturers freedom to design the devices according to their custom specifications, without bothering about the software. Android platform is dynamic, which means that the architecture also changes with newer versions. However, the core components remain more or less static. Fig. (**1**) depicts the Android software stack, consisting of four layers [6].

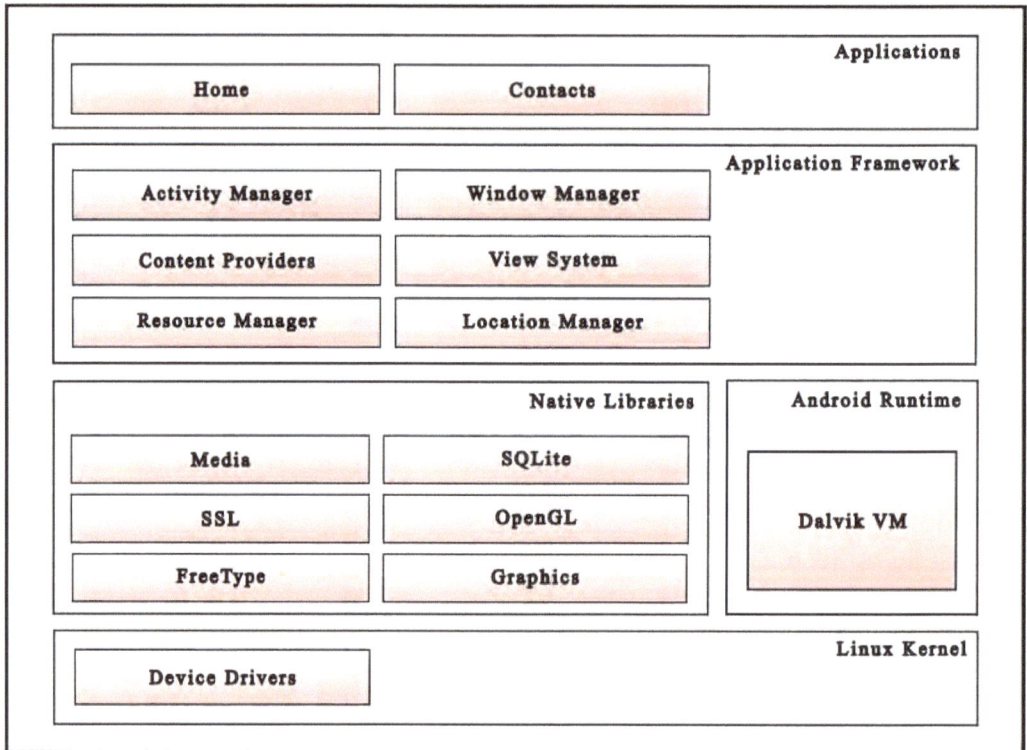

Fig. (1). Android Architecture [6].

Linux Kernel

Kernel acts as a hardware abstraction layer between hardware and other available software of the mobile device by providing a driver model. It is, therefore, responsible for memory management, process management, security model, networking, and other core OS services.

Platform Libraries and Android Runtime

On top of the Linux kernel, there are various function-specific libraries. It includes some popular C/C++- based libraries like libc, Webkit for browser compatibility, SSL for security, SQLite database for storage and sharing, media framework, *etc.* Java-based libraries specific to Android development are also present which support the application framework, and facilitate user interface

(UI), graphics, and database access, *e.g.*, Android, .app, .view, .os, .opengl, .widget, to name a few. This layer also includes the Android Runtime consisting of the Dalvik virtual machine (DVM; a specialised Java virtual machine for embedded systems, optimised to run in a resource-constrained environment). DVM uses multithreading and memory management features from Linux, to allow applications to create their processes. It allows application programmers to code using Standard Java [4].

Application Framework

Frameworks provide various types of interfaces, enabling the developer to develop apps without being concerned about the low-level details [6]. Key services included in this layer are activity manager, notification manager, views, resource manager, and content provider.

Applications

This is the top layer of the stack, where the apps are installed.

IOS PLATFORM

iOS is the proprietary operating system that powers mobile devices by Apple Corporation. Apple devices are in demand due to their highly acclaimed security features. Structurally its architecture is similar to the MacOS for computers. iOS comprises four layers, as shown in Fig. (**2**), each of which offers various special packages called frameworks [7]:

Core OS Layer

This layer is directly above the hardware and consists of basic frameworks that provide important OS services like kernel support, memory management, file management, resource management, access to external accessories, security, *etc.* Some of the important frameworks at this layer are security framework, networks framework, *etc.*

Core Services Layer

It primarily provides user data and accounts management through frameworks like Address Book framework, SQLite library, cloudkit framework, social framework, *etc.* It also provides location-related services through the location and motion frameworks. It also has a framework for managing in-app purchases.

Cocoa Touch (Application)
AppKit

Media

AV Manager	Core Animation	Core Audio	Core Image
Core Text	OpenAL	OpenGL	Quartz

Core Services

Address Book	Core Data	Core Foundation	Foundation
Quick Look	Social	Security	WebKit

Core OS

Accelerate	Directory Services	Disk Arbitration
	OpenCL	System Configuration	

Kernel and Device Drivers

BSD	File System	Mach	Networking

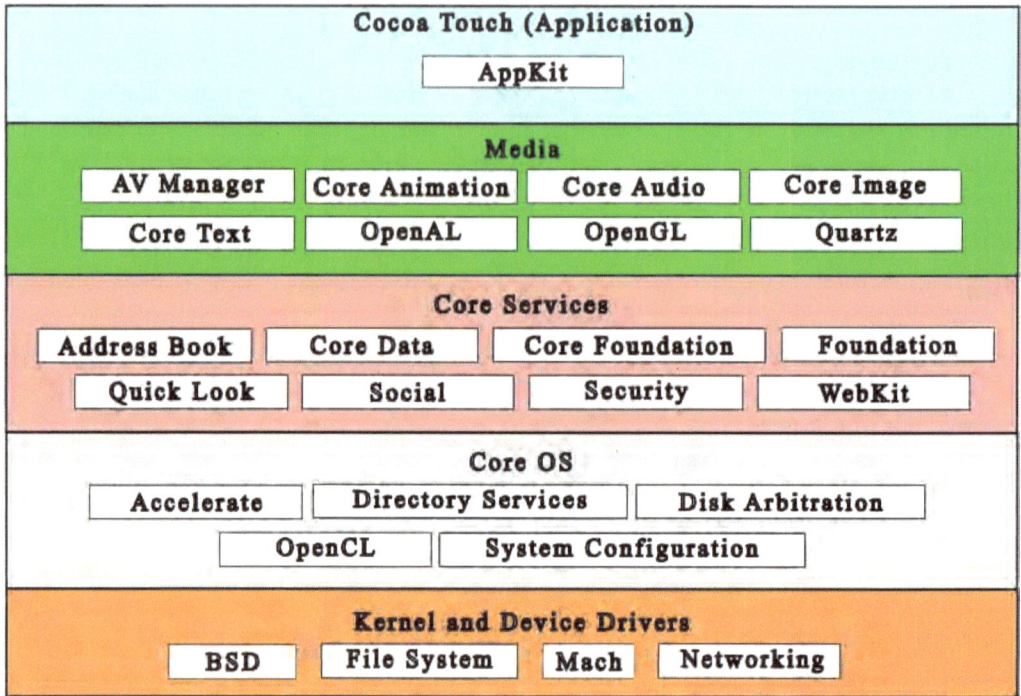

Fig. (2). iOS Architecture [7].

Media Layer

It provides multimedia support through various frameworks like media player framework for iTunes, core graphics framework, core animation framework, AV kit, *etc.*

Cocoa Touch

It is the top application layer providing user interface UI support via touch and gestures. It also provides registration services and notification management services through various frameworks. The frameworks in this layer are commonly used for app development.

DIGITAL FORENSICS

"The application of computer science and investigative procedures for a legal purpose, involving the analysis of digital evidence after proper search authority, chain of custody, validation with mathematics, use of validated tools, repeatability, reporting, and possible expert presentation" is referred to as digital forensics [8]. According to Palmer [9], a digital investigation consists of the following steps:

• Identification. It includes event or crime detection, identifying signatures, anomaly detection, system monitoring, audit analysis, *etc.*

• Preservation. Once the crime is detected, the case management is set. Next, the devices that could be a source of potential evidence are imaged and hashed, and the chain of custody is established. This happens in the second phase called preservation, which is a guarded principle throughout the digital forensic process.

• Collection. Approved methods, software, and hardware are used to collect and recover data. If necessary, lossless compression techniques are used.

• Examination and Analysis. Activities such as tracing the evidence, pattern matching, discovering and extracting hidden data and patterns, timelining, *etc.* are carried out on the bitwise copy of the evidence.

• Presentation. Final reporting, documentation, expert testimony, and so on.

The first three steps can be broadly categorized as "acquisition" and the last three steps as "analysis". This generic framework provides broad guidelines for conducting forensic investigation on any device, but the tools and techniques used in each of the steps can be specialized depending on the device or environment that is being investigated. Recent and upcoming computing environments can only be handled in a partial and fragmented manner using the existing toolkit in digital forensics and therefore they need to be treated as separate (though not necessarily independent) sub-classes of digital forensics. One such emerging area is that of mobile forensics. It includes both, the mobile device (hardware and software) as well as network connectivity.

The Need for Mobile Forensics as a Sub-Domain of Digital Forensics

Given the complete embedding of mobile devices in our day-to-day lives through a wide variety of apps and their ability to capture, store, and transmit user data and metadata, they are extremely valuable from a forensic perspective. The need for mobile forensics in day-to-day life is a matter of serious discussion.

Use of Mobile Phones to Store and Transmit Personal and Corporate Information

Mobile phones are widely used to send, receive, and share multimedia content. Mobile internet users use mobile phones for downloading music videos, web searching, instant messages, emails, various social networking sites, downloading ringtones, *etc.* The number of mobile internet users is estimated to have crossed 4 billion in 2019, and they constitute over 92% of the total population of active internet users. In India alone, 29% of the population access the internet from their mobile phones, which is expected to reach 35% by the year 2023 [3, 10]. Most enterprises allow their employees to bring their own devices (BYOD). In such a situation, smartphones may also have access to business information. The sensitivity of such information is determined by the position occupied by the individual; a more coveted position implies more sensitive information and, therefore higher value. Furthermore, financial and/or governance-related tasks like e-ballot, *etc.* are also carried out with smartphones. All this attracts the attention of hackers and cyber-criminals who try to exploit security loopholes and/or user behavior for their malicious activities.

Use of Mobile Phones in Online Transactions

Mobile phones enable easy online transactions through digital wallets (e-Wallets), mobile banking, internet banking, and payment interfaces. They are used for online shopping, payment of bills, investments, *etc.* Our accounts are linked to our mobile numbers. Cryptocurrencies are likely to make the usage of smartphones in financial systems even more prominent. Unfortunately, the involvement of money also attracts cyber-criminals, who use attacks varying in degree of sophistication, to catch the users unaware.

Mobile Phones as a Source of Big-data

Early mobile phones did not have the capacity to store large amounts of information. The only interesting information was the call records or device location obtained from the telecommunication companies with appropriate warrants. Lately, mobile phones have large storage capacity and a wide array of applications and connectivity options besides SIM cards from the network provider and the devices themselves have become forensically interesting.

Therefore, there is a dire need to treat mobile forensics differently than traditional forensics.

MOBILE FORENSICS

Framework

For digital investigations involving mobile devices, as in traditional devices, acquisition (identification, preservation, collection) and analysis (examination, analysis, presentation) need to be performed in a time-bound manner. However, acquisition in the case of mobile devices is trickier and often requires knowledge from multiple sub-domains of digital forensics. Once the evidence is acquired, analysis can be done using traditional tools and techniques.

Identification

It connotes finding out what is required for a particular case, which is to be found from the huge database of the mobile phone or the electronic device running on an Android system. Potential sources of evidence in a mobile phone include the SIM module, the device memory which includes volatile RAM and flash memory as well as SD cards, and more recently, mobile sensors like location, accelerometer, *etc.* Interesting artifacts may be found in the file system, device logs, app data stored in the mobile cloud, messenger apps, call records, *etc.*

Preservation

It suggests that the data present at a particular time should be protected from any external radio wave interference that is Wi-Fi, Blue- tooth, hotspot, VPN, *etc.* which can give access to the internet or outer world and alter the data of the device and change or alter the evidence needed. The ideal way to do this is to put the device in the Faraday bag so that the integrity of evidence can be kept. Seizing implies protecting the device from external alterations, which may be due to mishandling of evidence by electric shock, excessive heat, *etc.* Taking pictures and videos of the device, its make, manufacturer, serial number, operating system versions, *etc.* help in the acquisition phase of the whole investigation [11].

Collection

Extraction of data from mobile devices can be conducted in the following ways:

Manual Acquisition

The examiner uses the UI of the phone to browse and investigate. This is the least complex approach in terms of technical knowledge and there is no need for specialized tools. However, this approach has several drawbacks. Only the data that is visible through the UI of the device can be extracted, therefore deleted or hidden data may not be recovered with this technique. It is also unlikely to work with damaged devices. Moreover, there is a high risk of data modification as the investigator interacts with the device, and is time-consuming especially for devices with large memory.

Logical Acquisition

This is the process of extracting the files that are present on the file system partition, like call logs, SMS, browser history, people, contacts, images, *etc.* This technique uses the manufacturer's API's (Application Programming Interface) for synchronizing the phone contents. However, the security and forensic soundness of such APIs are questionable. The risk of inadvertent modification is still there and access to trace/remnant data could be limited.

Physical Extraction

This is also known as Hex dumping and is collected through standard interfaces like Joint Test Action Group (JTAG), USB, *etc.* It can be invasive and demands extensive skills in handling the extraction. It can extract data from devices with minor damage and can even retrieve remnant data, although it is difficult to parse and decode data because of non-standard formats.

Chip-Off

This technique involves extraction and imaging of the microchip. It is technically challenging and costly. If handled improperly, it can result in permanent loss of data.

Micro Read

This technique requires the use of an electron microscope in order to view and decode the information on memory chips by analyzing the physical circuits [12].

Examination and Analysis

Having collected the potential evidence, this step involves sifting, classification, and prioritizing data based on the case background. This step involves various activities for processing the data, drawing correlations, establishing a timeline, formulation of a hypothesis and statistical analysis, and finally preparing a case based on the results of the analysis.

Presentation

This includes reporting of the evidence collected in as simplified manner as possible, avoiding technical jargon, so that it is understood by the other stakeholders in the investigation and legal proceedings.

Tools

The discussion in the previous sections was on Android architecture and forensic framework suitable for mobile forensics. Further discussion will be on the software, which can retrieve the data from these layers. There are various tools available in the market to retrieve the data at various levels of acquisition: from simple data retrieval techniques of manual extraction, which mainly focuses on UI- based software, to micro read and chip off, where work is done at hardware level for data retrieval by bitwise decoding. It may be noted that the existing tools and techniques follow a hit-and-trial approach and rely largely on the technical ignorance of the device owner. Table **1** presents an overview of the popular tools used for data acquisition in mobile forensics.

Table 1. Data Acquisition Tools

Tool	Description	Manual	Logical Analysis	Physical Extraction	Chip off	Micro-read	Free and/or Open-Source	Android	iOS
XRY	Analyse and recover data from mobile devices	-	x	x	x	-	No	x	x
Oxygen Forensic Suite	Extract files, including metadata from mobile devices. Retrieval of cloud data and Alexa/google assistant connected IoT devices through the corresponding apps	-	x	x	-	-	No	x	x

(Table 1) cont.....

Tool	Description	Manual	Logical Analysis	Physical Extraction	Chip off	Micro-read	Free and/or Open-Source	Android	iOS
Lantern	For acquisition from Apple devices; multiple device acquisition into a single consolidated case file. Link analysis is automatically triggered and helps to decipher thousands of items of knowledge. File system viewer is integrated into the appliance for manual analysis with an inbuilt plist editor	-	x	x	-	-	No	x	x

(Table 1) cont.....

Tool	Description	Manual	Logical Analysis	Physical Extraction	Chip off	Micro-read	Free and/or Open-Source	Android	iOS
Cellebrite [13]	Cellebrite UFED supports a broad range of devices and over 30000 device profiles. It can bypass locks and even overcome encryption. It is available on multiple platforms.	-	x	x	-	-	No	x	x
	Cellebrite Physical Analyser supports over 40 languages and is a powerful decoding and parsing tool that supports multiple formats. It can retrieve deleted data as well as cloud data. It can even streamline the workflow to enhance cooperation between the stakeholders	-	x	x	x	-	No	x	x
iSesamo	Hardware tool for opening phone	-	-	-	x	-	No	x	x
Edec Eclipse	Captures video/images of device handling for the purpose of chain of custody	x	-	-	-	-	No	x	x

(Table 1) cont.....

Tool	Description	Manual	Logical Analysis	Physical Extraction	Chip off	Micro-read	Free and/or Open-Source	Android	iOS
Fernico ZRT	Captures video/images of device handling for the purpose of chain of custody	x	-	-	-	-	No	x	x
Pandora's Box	Tool for decoding Hex dumps	-	-	x	-	-	-	-	-
Andriller	Read-only, forensically sound, non-destructive acquisition from Android devices. It has features, such as powerful Lock screen cracking for Pattern, PIN code, or Password.	-	x	-	-	-	x	x	-
Fieta	Microscope	-	-	-	-	x	-	x	x
ADB	Android Debug Bridge is a command line tool for accessing Android devices via desktop	-	x	-	-	-	x	x	-
LiME	Linux memory extractor	-	x	x	-	-	x	x	-

Recent Developments

Table **2** presents an overview of the literature in mobile forensics since 2011.

Table 2. Recent Developments

Paper	Contribution
[14]	Case study for physical and logical acquisition from Sony Xperia 10i
[15]	Framework for Android Forensics
[16]	Collection methodology
[17]	Geo Data analysis by leveraging app permissions
[18]	Data recovery and analysis of fragmented flash memory
[19]	Method and tools for volatile memory acquisition
[20]	Data collection through software package
[21]	Analysis of raw NAND flash memory
[22]	Adversary model
[23]	Methodology for collection of evidence: Start with bootloader and live acquisition followed by complete imaging.
[24]	Data perspective on Android forensics in terms of orientation, structure, extraction, and analysis.
[25]	Volatile memory data capturing using the system-level data migration function offered by Android and a forensically capable intermediate device

Challenges

Issues related to Process Models

Extensive testing and verification across the wide technological landscape to prove that the suggested high-level process flow is indeed practical, forensically sound, and generally applicable.

Tool Development

Existing mobile forensics tools come with poor documentation; impairing explainability and raising questions about integrity. Furthermore, it has been observed that evidence extraction is tool-dependent. Using different tools on the same dataset may sometimes result in disparate, possibly contradictory outcomes. The soundness of forensic tools may be undermined due to bugs and vulnerabilities, which is why there is an emphasis on the verification of tools in digital forensics. However, in the case of mobile forensics tools, by the time the verification process is completed, it is likely that the tool in its erstwhile version would have become obsolete. Thus, there is a need to focus on upgrading the tool verification infrastructure.

Problems due to Software Stack in Mobile Devices

Poor documentation in apps and less popular/legacy operating systems leads to frequent dependence on reverse engineering. Furthermore, often specialized APIs are needed for extracting evidence from the various apps. Things are dependent on the technical capability of the forensic investigator due to the heterogeneity of formats across the apps, which affects the parsing and interpretation of evidence. Furthermore, frequent application and platform upgradation and new releases can cause a validation crisis for previously verified tools and techniques, possibly rendering them obsolete. Customized as well as counterfeit device configurations pose additional challenges for the investigators.

Technological Evolution

The evolution of mobile computing has been pushed by cloud processing, since almost all apps use cloud services at least for storage and backup of user data, keeping track of user logs and metadata and so on. Evidence extraction from the cloud poses additional problems of its own, like, jurisdiction issues, multi-tenancy, data ownership and location of evidence in a virtual and distributed environment, and so on. On the other hand, IoT integration to mobile devices is leading to new problems with respect to evidence extraction.

Problems with Big Data Volume, Volatility, Variety

In mobile devices, there is a significant amount of data at multiple levels. Local data is stored in flash memory and RAM. A lot of data is in transit on the network and interesting data may also be stored in the cloud. Therefore, these devices present the problem of dealing with large amounts of volatile data, creating challenges for device isolation and maintaining/proving evidence integrity.

Security Features and Anti-forensics

The built-in security features in devices and applications pose significant challenges to the acquisition and evidence extraction process. *E.g.*, Encryption for protecting user data creates serious roadblocks for the investigation, not easily overcome by existing forensic tools. Furthermore, advanced users may use custom security features and anti-forensic techniques like secure erase, steganography, app mutations, *etc.* in order to derail evidence extraction, which are often difficult to detect. The presence of malware may further complicate the process, not only posing risks for evidence integrity, but possibly compromising the forensic workstation being used for extraction.

Miscellaneous Non-technical Issues

Operational and Personnel issues. Ethical and legal issues.

Opportunities

Upgradation of Toolkit

Existing process models and tools/techniques need to be upgraded and new ones developed in order to deal with the challenges discussed in the previous subsection. They should be independent of architectural differences, holistic, proactive, intelligent and scalable for future technological changes. Furthermore, there is a need to rationalize and revamp the validation process and verification processes for forensic soundness.

Automation

In order to reduce the dependence on app specific APIs, there is a need to evolve generic automated approaches to data extraction. These will be useful for rapid extraction and parsing of data from the device and its contextualization using cloud data.

Intelligent Analysis

Researchers are actively exploring the use of big data analytics and artificial intelligence techniques for dealing with the problems of big data in mobile forensics and for improving presentation.

Training and Skill Development

Investigators and other law-enforcement personnel need to be trained and re-skilled from time to time in order to keep up with technological changes and tool development. Thus, they need to work in close association with industry and academia and evolve an ecosystem for training and awareness.

CONCLUSION

Mobile devices are ubiquitous. The proximity of these devices to the owner not only make them sources of valuable information but also their owners' digital witnesses. The nature, architecture and construction of mobile devices demands new frameworks and techniques for their forensic analysis, as traditional techniques are unable to handle the challenges posed by these devices.

This chapter provided an overview of mobile forensics from the perspective of data acquisition and briefly presented the framework, tools and techniques used for the purpose. It also presented a list of notable works in mobile forensic data acquisition. Furthermore, the challenges and opportunities in mobile forensics were identified, which point towards the hot research areas in the subject. Standardization and validation of procedures, tools and techniques for mobile forensics is an imminent need. Automation is another area that requires attention, most of the evidence acquisition still happens manually.

REFERENCES

[1]   C. D Orazio, A. Ariffin, A. and K.K.R. Choo, "IoS anti-forensics: How can we securely conceal, delete and insert data?", *2014 47th Hawaii International Conference on System Sciences,* pp. 4838-4847, 2014.
       [http://dx.doi.org/10.1109/HICSS.2014.594]

[2]   K. Barmpatsalou, T. Cruz, T., E. Monteiro and P. Simoes, "Current and future trends in mobile device forensics: A survey", *ACM Comput. Surv,* vol. 51, no. 3, 2018.
       [http://dx.doi.org/10.1145/3177847]

[3]   Available from: www.statista.com/statistics/272698/global-market-share-held-by-mobile-operating systems-since-2009/

[4]   C. Racioppo, and N. Murthy, "Android forensics: A case study of the 'HTCincredible 'phone", In: *Proceedings of Student-Faculty Research Day*, 2012, pp. 1-8.

[5]   A. Hoog, "Android forensics: investigation, analysis and mobile security for Google Android", *Elsevier,* 2011.
       [http://dx.doi.org/10.1016/B978-1-59749-651-3.10001-9]

[6]   V. Vijayan, Android forensic capability and evaluation of extraction tools, Master's thesis, Edinburgh Napier University, 2012. Unpublished.

[7]   S. Chauhan, "Understanding Xamarin IOS - build native IOS app," Live Training, Prepare for Interviews, and Get Hired, Available from: https://www.dotnettricks.com/learn/xamarin/understanding-xamarin-ios-build-native-ios-app

[8]   K. Zatyko, "Commentary: Defining digital forensics", 2007.

[9]   G. Palmer, "A road map for digital forensics research report from the first digital forensics research workshop (dfrws), dtrt001–01", *Tech. rep,* 2001.

[10] Available from: https://www.statista.com/statistics/617136/digital-population-worldwide/

[11] M. Faheem, M.T. Kechadi, and N.A. Le-Khac, The state-of-the-art forensic techniques in mobile cloud environment: A survey, challenges and current trends.*Mobile computing and wireless networks: Concepts, methodologies, tools, and ap- plications.* IGI Global, 2016, pp. 2111-2131.
       [http://dx.doi.org/10.4018/978-1-4666-8751-6.ch092]

[12] R. Ayers, S. Brothers, and W. Jansen, "Guidelines on mobile device forensics", 2014.
       [http://dx.doi.org/10.6028/NIST.SP.800-101r1]

[13] Available from: https://www.cellebrite.com/en/product/

[14] D. Quick, and M. Alzaabi, ""Forensic analysis of the Android le system YAFFS2", In: *Cowan Univ* Tech. Rep: Joondalup, WA, Australia, 2011, pp. 100-109.

[15] A. M. L. de Simão, F. C. Sícoli, L. P. de Melo, F. E. G. de Deus, and R. T. de Sousa Júnior, "Acquisition and analysis of digital evidence in Android smartphones", *Tech. Rep,* 2011.

[16] T. Vidas, C. Zhang, and N. Christin, "Toward a general collection methodology for Android devices", *Digit. Invest,* vol. 8, pp. S14-S24, 2011.
[http://dx.doi.org/10.1016/j.diin.2011.05.003]

[17] S. Maus, H. Höfken, and M. Schuba, "Forensic analysis of geodata in Android smartphones", *Tech. Rep.,* 2011.

[18] J. Park, H. Chung, and S. Lee, "Forensic analysis techniques for fragmented flash memory pages in smartphones", *Digit. Invest,* vol. 9, no. 2, pp. 109-118, 2012.
[http://dx.doi.org/10.1016/j.diin.2012.09.003]

[19] J. Sylve, A. Case, L. Marziale, and G. G. Richard, "Acquisition and analysis of volatile memory from Android devices", *Digit. Invest,* vol. 8, no. 3, pp. 175-184, 2012.
[http://dx.doi.org/10.1016/j.diin.2011.10.003]

[20] D. Votipka, T. Vidas, and N. Christin, "Passe-Partout: A General Collection Methodology for Android Devices", *IEEE transaction on information forensics and security,* vol. 8, no. 12, 2013.

[21] D. B. L. Schatz, "A visual approach to interpreting NAND flash memory", *Digit. Invest,* vol. 11, no. 3, pp. 214-223, 2014.
[http://dx.doi.org/10.1016/j.diin.2014.05.018]

[22] Q. Do, B. Martini, and K.K.R. Choo, "A forensically sound adversary model for mobile devices", *PLoS One,* vol. 10, no. 9, p. e0138449, 2015.
[http://dx.doi.org/10.1371/journal.pone.0138449] [PMID: 26393812]

[23] B. Martini, Q. Do, and K.K.R. Choo, *Conceptual Evidence Collection and Analysis Methodology for Android Devices.* Cloud Security Ecosystem. Syngress, 2015, pp. 285-307.
[http://dx.doi.org/10.1016/B978-0-12-801595-7.00014-8]

[24] N. Scrivens, and X. Lin, "Android Digital Forensics: Data, Extraction and Analysis", *Proceedings of the ACM Turing 50th Celebration Conference-China,* 2017.
[http://dx.doi.org/10.1145/3063955.3063981]

[25] P. Feng, Q. Li, P. Zhang, and Z. Chen, "Private data acquisition method based on system-level data migration and volatile memory forensics for android applications", *IEEE Access,* vol. 7, 2019.

CHAPTER 9

IoT and AIoT: Applications, Challenges and Optimization

Amit Verma[1,*] and **Raman Kumar**[2]

[1] Maharaja Agrasen Institute of Technology, Maharaja Agrasen University, Himachal Pradesh, India

[2] IKGPTU, Kapurthala, Jalandhar, Punjab, India

Abstract: The Internet of Things (IoT) has rapidly gained popularity as a technology that enables devices to communicate with each other and the Internet, opening up a world of possibilities for new applications and services. This chapter provides an overview of IoT, its applications, and the challenges that need to be addressed in its deployment. IoT and AIoT are two of the most significant technological innovations of the 21st century. IoT allows physical devices to connect and exchange data, while AIoT enables these devices to learn, analyze, and make decisions based on the data they collect. The term "AIOT" stands for "Artificial Intelligence of Things." AIOT refers to the integration of Artificial Intelligence (AI) technologies with the Internet of Things (IoT) ecosystem. In essence, AIOT combines the capabilities of AI and IoT to create intelligent, self-learning systems that can analyze, interpret, and respond to data generated by IoT devices. Together, these technologies offer numerous benefits such as increased efficiency, better decision-making capabilities, and improved outcomes across industries like healthcare, manufacturing, transportation, and agriculture. As more devices and systems become connected, IoT and AIoT will continue to play a critical role in shaping the future of our world. IoT and AIoT have the potential to transform the way we live and work. By enabling devices to communicate and share data, IoT can help us create more efficient and effective systems, and by integrating AI technologies, IoT devices can become smarter and more autonomous. This means that devices can analyze data in real-time, make decisions, and adapt to changing conditions without human intervention. For example, in smart cities, IoT and AIoT can help reduce traffic congestion by optimizing traffic flows, and in healthcare, they can help monitor patients remotely and alert healthcare providers when necessary. As more devices and systems become connected, we can expect IoT and AIoT to become increasingly sophisticated, offering new opportunities for innovation and growth in various industries. However, as with any new technology, there are also potential risks and challenges that must be addressed, such as security and privacy concerns, and the need for new regulations and standards to ensure the safe and ethical use of these technologies.

---

[*] **Corresponding author Amit Verma:** Maharaja Agrasen Institute of Technology, Maharaja Agrasen University, Himachal Pradesh, India; E-mail: verma0152@gmail.com

Neha Kishore, Pankaj Nanglia, Shilpa Gupta & Ashutosh Kumar Dubey (Eds.)

Keywords: Artificial intelligence, AI, AIoT, Applications, Challenges, Internet of things, Optimization, IoT.

INTRODUCTION

The Internet of Things (IoT) is a term used to describe the network of physical devices, appliances, and industrial equipment that are connected to the internet and can communicate with each other. These devices are equipped with sensors and software that enable them to collect and share data, facilitating remote monitoring and control as well as automated decision-making. With IoT, centralized systems can gather data from multiple sources and use it to optimize performance, reduce costs, and improve efficiency. This technology has applications in various fields, such as healthcare, transportation, and smart homes, and has the potential to create smart cities that can improve the quality of life and sustainability. The advent of this technology has the capability to transform a number of industries and enhance our daily lives in many ways [1].

• One of the most significant benefits of IoT is the ability to collect and analyze real-time data. This allows for a wide range of applications, such as predictive maintenance, remote monitoring, and automated decision-making. For example, in the industrial sector, IoT devices can be used to monitor and control industrial equipment, enabling predictive maintenance and increasing efficiency [2, 3]. This can lead to significant cost savings for companies and reduce downtime as shown in Fig. (**1**).

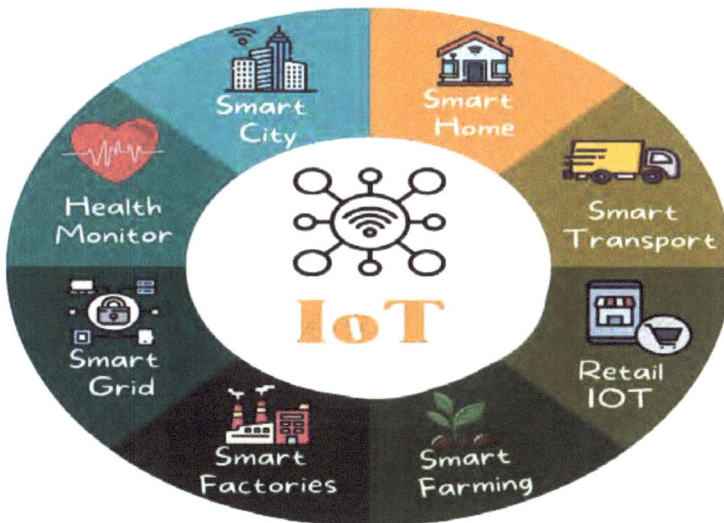

Fig. (1). IoT applications.

• Another major benefit of IoT is the ability to create smart cities and homes. IoT devices can be used to monitor and manage traffic, energy usage, and other aspects of urban infrastructure, making cities more efficient and sustainable. In homes, IoT devices can be used to control lighting, temperature, and appliances, making it possible to create a truly connected and automated living space [4]. This can lead to increased comfort, convenience, and energy savings for homeowners.

• IoT devices have numerous applications in healthcare [5]. They can monitor patients' health and transmit real-time data to healthcare professionals, enabling more precise diagnoses and treatments. Wearable health monitors, for example, can track vital signs and provide doctors with health information from remote locations. This has the potential to enhance patient outcomes and decrease healthcare expenses.

• The use of IoT technology also has the ability to enhance logistics and supply chain management. IoT devices can track the movement of goods, monitor inventory levels, and optimize delivery routes, among other things. This can lead to higher efficiency, lower expenses, and improved customer service.

• IoT technology has the potential to revolutionize many industries and improve the way we live our lives [4]. However, it is important to note that IoT devices are also vulnerable to cyber-attacks. With more connected devices, the risk of cyber-attacks increases. This is an important concern that needs to be addressed. Companies and individuals must take steps to secure their IoT devices and protect their data from cybercriminals [6, 7].

There are several techniques and technologies used in IoT to connect devices and gather data. Some of the most common include:

Wireless communication: IoT devices use wireless technologies like Wi-Fi, Bluetooth, Zigbee, and cellular networks to communicate with one another as well as centralized systems, facilitating easy device connectivity and remote control [8].

Sensors: IoT devices use sensors to collect data on their environment, such as temperature, humidity, and motion. This data can be used to make automated decisions and control other devices.

Cloud computing: IoT devices often use cloud computing services to store and process data. This allows for easy data analysis and sharing, and enables devices to communicate with each other regardless of location [8].

Big data and analytics: IoT generates a huge amount of data, which can be used to improve efficiency, decision making and even predict future outcomes. To make sense of this data, IoT systems often use big data and analytics tools to extract insights and make predictions.

Artificial intelligence and machine learning: IoT devices can use artificial intelligence and machine learning algorithms to analyze data, make predictions, and automate decision-making. This allows for intelligent and responsive control of devices [9].

Block chain: This technique is used to secure the data and devices for IoT. Block chain can be used to protect against cyber-attacks, ensure data integrity and authenticity, and enable secure communication and data sharing among devices.

These are some of the most common techniques used in IoT to connect devices and gather data. However, there are many other technologies and approaches being developed and used in the field, and it continues to evolve.

IoT Networks: IoT networks refer to the infrastructure and technology that enables the communication and connectivity of Internet of Things (IoT) devices [10]. These networks consist of a variety of components, including:

• **IoT devices:** These are the physical objects that are connected to the internet and equipped with sensors, actuators, and communication capabilities. Examples include smart home devices, industrial equipment, and healthcare devices.

• **Gateways:** These are devices that act as a bridge between IoT devices and the internet. They collect data from IoT devices, process it, and send it to the cloud or other centralized systems for storage and analysis.

• **Network infrastructure:** This includes the wireless and wired communication networks that connect IoT devices to the internet, such as Wi-Fi, cellular, and LoRa networks.

• **Cloud computing:** IoT networks often use cloud computing services to store and process data, allowing for easy data analysis and sharing.

• **Platforms and applications:** These are the software systems and applications that run on top of the IoT network infrastructure and provide the functionality and user interfaces for interacting with IoT devices.

• **Security:** When developing IoT networks, it is critical to prioritize security in order to safeguard against cyber-attacks and unauthorized access. This involves implementing encryption, firewalls, and other security measures.

• IoT networks are complex and varied, and can be designed for specific use cases and industries. The design of IoT networks will depend on the specific requirements of the IoT application and the devices that will be connected.

Overall, IoT networks are the backbone of IoT technology, providing the communication and connectivity infrastructure that enables the collection and sharing of data, and enables IoT devices to communicate with each other and with centralized systems.

CHALLENGES IN IOT

The Internet of Things (IoT) networks face various issues, including:

• **Security:** One of the major issues in IoT networks is security. IoT devices are vulnerable to cyber-attacks and can be easily compromised. This can lead to data breaches and unauthorized access to devices, putting both individuals and organizations at risk [11].

• **Interoperability:** IoT networks often consist of a wide range of devices from different manufacturers, which can lead to interoperability issues. In some cases, devices may not have the ability to communicate with each other or share data in a standardized format.

• **Scalability:** IoT networks can generate a large amount of data and require significant processing power to analyze it. This can be a challenge for IoT systems that need to scale to accommodate large numbers of devices and users [12].

• **Reliability:** IoT networks must be highly reliable to ensure that devices can communicate and share data without interruption. This can be challenging as IoT networks often rely on wireless communication and can be affected by interference and other factors [13].

• **Privacy:** With the increasing amount of data being generated and collected by IoT devices, there is a growing concern about privacy. Individuals may be concerned about how their data is being used and shared, and organizations must ensure that they are complying with privacy regulations [14].

• **Energy consumption:** IoT devices are generally small and portable, but they require energy to work. This can be a problem when devices are deployed in remote locations or in situations where power is limited [15].

• **Complexity:** With the increasing number of devices and the complexity of the systems that they are connected to, it can be difficult to manage and maintain IoT networks. This can be a challenge for organizations that need to ensure that their devices and systems are working properly.

These are some of the major issues facing IoT networks. Addressing these issues is critical to ensure that IoT networks are secure, reliable, and able to accommodate the growing number of devices and users.

OPTIMIZATION IN IOT NETWORKS

Optimization in IoT networks is critical to ensure that the network is performing efficiently and effectively. Here are some ways to optimize IoT networks:

• **Data compression** is a technique used to reduce the amount of data that is transmitted over the network. In IoT networks, data is generated by a large number of sensors and devices, and this data can consume significant bandwidth when transmitted over the network. Data compression can help to reduce the amount of data transmitted, which can improve network performance and reduce the costs associated with transmitting and storing data [16]. Data compression works by encoding data in a more efficient format that requires fewer bits to represent the same information. There are several techniques used for data compression, including lossless and lossy compression.

Lossless compression algorithms preserve all of the original data when compressing it, ensuring that the original data can be perfectly reconstructed when decompressed. This type of compression is ideal for data that cannot be modified or changed, such as sensor readings or historical data. Lossless compression algorithms include Huffman coding, Lempel-Ziv-Welch (LZW) coding, and arithmetic coding.

Lossy compression algorithms, on the other hand, sacrifice some of the original data when compressing it, resulting in a smaller compressed file. This type of compression is ideal for data that can be modified or changed without significant impact, such as images or audio. Lossy compression algorithms include JPEG for images, MP3 for audio, and MPEG for video.

When implementing data compression in IoT networks, it is important to consider the trade-off between the amount of compression and the impact on the data quality. Lossless compression is generally preferred for data that cannot be modified, while lossy compression can be used for data that can be modified with a minimal impact on its usefulness. Overall, data compression is an important optimization technique for IoT networks, as it can reduce the amount of data transmitted over the network and improve network performance.

• **Load balancing** is a technique used to distribute network traffic across multiple servers or devices to prevent any one server or device from becoming overloaded. In IoT networks, load balancing can be used to ensure that network resources are used efficiently and effectively, and to prevent any one device or server from becoming a bottleneck.

There are multiple ways to implement load balancing. One common method is to use a load balancer, which is a software application or device that distributes network traffic across multiple servers or devices. The load balancer can use different algorithms to distribute traffic, including round-robin, least connections, or IP hash. Another option is distributed load balancing, which involves distributing the load balancing function across multiple devices or servers. In this approach, each device or server can manage a portion of the network traffic [7] [9] and can distribute it to other devices or servers as required.

Load balancing can help to improve network performance and availability in several ways. By distributing network traffic across multiple devices or servers, load balancing can prevent any one device or server from becoming overloaded, which can reduce latency and improve response times. Load balancing can also help to ensure that network resources are used efficiently and effectively, which can reduce the cost of maintaining and scaling the network.

In IoT networks, load balancing can be particularly important due to the large number of devices and sensors that generate data. By using load balancing, network administrators can ensure that network resources are used efficiently and effectively and that data is transmitted and processed in a timely and reliable manner.

• **Quality of Service (QoS)** refers to the ability of a network to deliver different types of traffic with varying requirements for reliability, bandwidth, latency, and other performance metrics. In IoT networks, QoS is essential because IoT devices typically transmit different types of data with different priority levels and characteristics. QoS in IoT networks involves several aspects such as network architecture, protocol design, traffic engineering [9], and resource management. The goal of QoS is to ensure that IoT devices and applications receive the required level of service while minimizing the impact on network resources and other applications.

One of the primary challenges in QoS for IoT networks is managing the heterogeneous traffic from different types of devices with varying requirements. For example, some IoT devices may require high-bandwidth and low-latency connections, while others may have lower bandwidth requirements but higher reliability needs. QoS mechanisms such as traffic shaping, packet prioritization and congestion control can be used to manage this traffic.

Another aspect of QoS in IoT networks is security and privacy. Since IoT devices may transmit sensitive data, such as health data or financial information, QoS mechanisms should ensure that the data is transmitted securely and with high reliability. In summary, QoS is an essential aspect of IoT networks, and it ensures that the network can deliver different types of traffic with varying requirements for reliability, bandwidth, latency, and security. QoS mechanisms should be designed to manage the heterogeneous traffic from IoT devices while minimizing the impact on network resources and other applications.

AIoT (Artificial Intelligence of Things)

Intelligent IoT, also called AIoT (Artificial Intelligence of Things), is the integration of artificial intelligence (AI) technologies with IoT devices [11]. These devices are connected to the internet and gather data from sensors, cameras, and other sources. These devices generate a massive amount of data, which can be analyzed and processed to extract valuable insights that can be used to improve the efficiency of business operations, enhance customer experience, and create new business opportunities [12, 24].

Integrating AI with IoT devices enables the devices to learn and adapt to changing conditions in real time. AI algorithms can analyze data collected from IoT devices and provide insights into the behavior and patterns of the devices. This can be used to optimize device performance, predict failures, and improve device maintenance [13].

Intelligent IoT can also enable devices to make decisions autonomously without human intervention. Overall, the combination of AI and IoT can create intelligent systems that are more efficient, more responsive, and more adaptable to changing conditions [14]. This can lead to significant improvements in productivity, cost savings, and overall quality of life. AI-based IoT involves the use of AI technologies such as machine learning (ML), natural language processing (NLP), and computer vision to analyze and interpret the massive amounts of data generated by IoT devices [25]. These technologies enable IoT devices to become more intelligent [15], adaptive, and capable of making decisions without human intervention.

Some examples of AI-based IoT applications as shown in Fig. (2) are briefly discussed below:

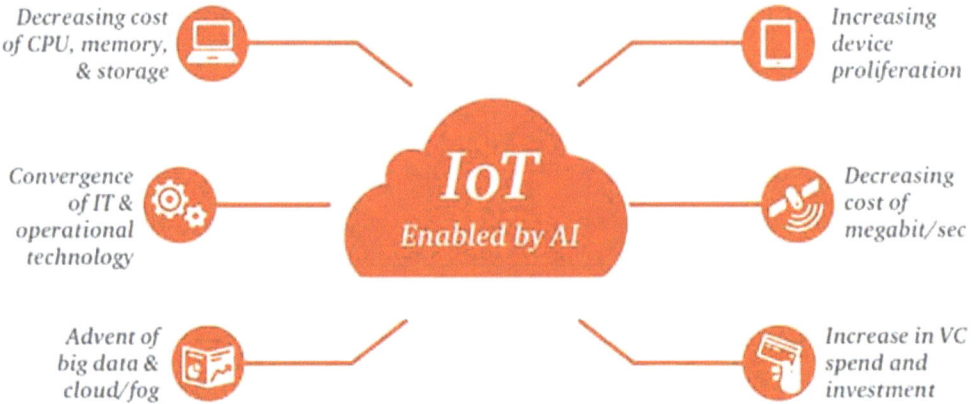

Fig. (2). AIoT (Artificial Intelligence of Things).

Predictive maintenance: Through the analysis of data gathered from sensors and other sources, AI algorithms can predict when equipment or machines are likely to fail, enabling proactive maintenance.

Smart energy management: AI can be used to analyze energy usage patterns and optimize energy consumption, reducing costs and improving sustainability.

Autonomous vehicles: Self-driving cars rely on AI to analyze real-time traffic data and make decisions regarding driving behavior [8].

Smart home automation: AI-powered smart home devices can learn user preferences and adjust lighting, temperature, and other settings accordingly.

Healthcare monitoring: Wearable IoT devices can collect data on patient health and use AI to detect abnormalities or changes that may require medical attention [6].

AI-based IoT can also help organizations make better decisions by providing insights into customer behavior, operational efficiency, and product performance. For example, retailers can use AI-powered IoT devices to analyze customer behavior and adjust pricing and product offerings accordingly.

However, there are also challenges to implementing AI-based IoT systems, including data privacy and security concerns, the need for specialized expertise, and the potential for bias in AI algorithms. As such, it is important to consider these issues and take steps to address them when developing AI-based IoT solutions.

AIOT CHALLENGES

AIoT technologies face several challenges that must be addressed to ensure their safe and ethical use [18, 19]. Here are some of the most significant challenges:

• Data Privacy and Security: AIoT devices collect and store vast amounts of sensitive data, including personal and financial information, making them attractive targets for cybercriminals. Ensuring data privacy and security is critical to maintaining trust in AIoT technologies.

• Lack of Standardization: There is currently no widely accepted set of standards for AIoT devices, which can lead to compatibility issues and interoperability problems between devices from different manufacturers.

• Ethical Concerns: AIoT technologies raise ethical concerns related to data ownership, transparency, and accountability. The use of AIoT in areas such as surveillance, monitoring, and decision-making also raises questions about privacy, fairness, and bias.

• Complexity and Integration: The integration of AI and IoT can create complex systems that are difficult to manage and maintain. This can lead to operational issues and increase the risk of system failures [17 - 19].

• Limited Computing Resources: Many IoT devices have limited computing resources, which can make it challenging to run AI algorithms and perform data analysis. This can limit the effectiveness of AIoT devices in some use cases.

Overall, addressing these challenges is essential to ensure the safe and ethical use of AIoT technologies. As the use of AIoT becomes more widespread, it will be essential to develop new standards and best practices to address these challenges and unlock the full potential of these technologies [20].

CONCLUSION

In summary, the Internet of Things (IoT) is a game-changing technology with the potential to transform several industries and domains. However, the deployment of IoT systems comes with a host of challenges such as security, privacy, scalability, interoperability, and power consumption. By integrating artificial intelligence (AI) technologies, commonly referred to as AIoT, it is possible to address some of these challenges and optimize IoT networks. AIoT can enhance the capabilities of IoT devices by enabling them to learn from data, make autonomous decisions, and improve their performance over time. The integration of AI with IoT can lead to the development of more intelligent and adaptive systems that can bring significant benefits to businesses and society as a whole.

However, the integration of AI with IoT also presents challenges such as data privacy, security, and the complexity of AI algorithms [23]. To ensure the successful deployment of AIoT systems, these challenges must be addressed. This chapter provides an overview of IoT, its applications, and the challenges associated with its deployment. It also discusses the potential benefits and challenges of integrating AI technologies with IoT networks to create AIoT systems. We hope that this chapter will serve as a valuable resource for researchers, practitioners, and students interested in understanding the latest developments in IoT and AIoT [22]. The future scope of IoT and AIoT is vast, and we can expect both technologies to continue to evolve and become more advanced in the years to come. With the rise of edge computing, blockchain integration, 5G networks, cognitive IoT, and autonomous systems, we will see more efficient and intelligent devices capable of making better decisions in real time. These developments will have a significant impact on various industries, including healthcare, manufacturing, transportation, and agriculture, among others. As IoT and AIoT continue to mature, we can expect to see new use cases and applications emerge, offering endless opportunities for innovation and growth. However, as with any new technology, there are also potential risks and challenges that must be addressed to ensure the safe and ethical use of these technologies.

REFERENCES

[1]   S.R. Sarma, and S.K. Sahoo, "IoT-based smart home: A review", *IEEE 5th International Conference on Computing Communication Control and Automation (ICCUBEA), Pune, India,* pp. 1-5, 2019.
         [http://dx.doi.org/10.1109/ICCUBEA48708.2019.8970987]

[2]   T.H. Siddique, and K. Al-Ali, "IoT Security: A Review", *2020 4th International Conference on Intelligent Computing and Control Systems (ICICCS), Madurai, India,* pp. 196-200, 2020.
         [http://dx.doi.org/10.1109/ICICCS48743.2020.9123957]

[3]   A.W.F. Shalaby, S.A. Attia, A.S.A. Mahmoud, and A.A.K. Ibrahim, "IoT-Based Air Quality Monitoring System Using Machine Learning Techniques", *2021 International Conference on Innovative Trends in Information Technology (ICITIT),* pp. 1-6, 2021. Assiut, Egypt

[4]   Y. Sun, Y. Liu, X. Zhang, and Y. Lu, "An Efficient and Reliable IoT Data Management System for Smart Cities", *IEEE Internet Things J.,* vol. 6, no. 2, pp. 1763-1773, 2019.
         [http://dx.doi.org/10.1109/JIOT.2018.2889753]

[5]   F. Raza, A. Iqbal, M. Awais, and M.F. Bari, "Internet of Things (IoT) Based Healthcare System: A Review", *IEEE Access,* vol. 7, pp. 163779-163802, 2019.
         [http://dx.doi.org/10.1109/ACCESS.2019.2954025]

[6]   R. Pandey, N.N. Singh, and N.K. Sharma, "A Survey on IoT Applications in Agriculture", *2018 International Conference on Advances in Computing, Communications and Informatics (ICACCI),* pp. 2115-2119, 2018. Bangalore, India

[7]   S. Kumar, and N. Saxena, "A Review of IoT Based Traffic Management System", *2020 International Conference on Smart Cities and Green ICT Systems (SMARTGREENS),* pp. 77-83, 2020. Prague, Czech Republic

[8]   D.D.D.T. Dang, Q.A. Le, and H.C. Nguyen, "Smart Parking System Using IoT Technologies: A

Review", *IEEE Access,* vol. 7, pp. 19134-19144, 2019.
[http://dx.doi.org/10.1109/ACCESS.2019.2895216]

[9]   M.A. Naeem, M. Shahid, M. Imran, and M.T. Khan, "A Survey on Fog Computing for the Internet of Things: Current Trends and Future Directions", *IEEE Access,* vol. 7, pp. 79179-79207, 2019.
[http://dx.doi.org/10.1109/ACCESS.2019.2923147]

[10]   N. Patil, and K. Varshney, "IoT-Based Water Quality Monitoring System Using Wireless Sensor Network", 2019.

[11]   M.A. Malik, A. Bashir, M. Tariq, and A. Qadir, "Artificial Intelligence of Things (AIoT): A Comprehensive Survey", *IEEE Access,* vol. 8, pp. 129182-129211, 2020.
[http://dx.doi.org/10.1109/ACCESS.2020.3002188]

[12]   W. Gao, Z. Ma, Y. Xu, and X. Lu, "Design and Implementation of an AIoT System for Smart Home", *IEEE Trans. Consum. Electron.,* vol. 66, no. 1, pp. 73-80, 2020.
[http://dx.doi.org/10.1109/TCE.2020.2960404]

[13]   S.S. Das, and S. Mukhopadhyay, "AIoT: The Convergence of Artificial Intelligence with the Internet of Things", *IEEE Consum. Electron. Mag.,* vol. 10, no. 1, pp. 46-53, 2021.
[http://dx.doi.org/10.1109/MCE.2020.3039521]

[14]   R.M. Devarasetty, P. Manoharan, and J.B. Song, "Edge Intelligence for Industrial IoT: A Comprehensive Review", *IEEE Access,* vol. 9, pp. 1472-1492, 2021.
[http://dx.doi.org/10.1109/ACCESS.2020.3041681]

[15]   D.K.T. Nguyen, S. Choo, and K. Kim, "A Comprehensive Survey on AIoT for Smart Cities: Architecture, Applications, and Challenges", *IEEE Internet Things J.,* vol. 8, no. 2, pp. 915-935, 2021.
[http://dx.doi.org/10.1109/JIOT.2020.3033129]

[16]   Avialable from: https://ieeexplore.ieee.org/document/7468228

[17]   Avialable from: https://www.sciencedirect.com/science/article/pii/S2212671620303014

[18]   Avialable from: https://ieeexplore.ieee.org/document/7552094

[19]   Avialable from: https://www.sciencedirect.com/science/article/pii/S2405452620305566

[20]   Avialable from: https://www.sciencedirect.com/science/article/pii/S2212671620306469

[21]   A. Jara, M.A. Zamora, and A.F. Skarmeta, "The Internet of Things: A survey", *Comput. Netw.,* 2014.
[http://dx.doi.org/10.1016/j.comnet.2014.02.008]

[22]   B. Guo, W. Li, H. Song, B. Hu, and N. Liu, "Artificial Intelligence of Things (AIoT): A Comprehensive Survey", *Future Gener. Comput. Syst.,* 2018.
[http://dx.doi.org/10.1016/j.future.2017.12.055]

[23]   R, E., Varathan, K. "Internet of Things (IoT) and its impact on supply chain: A framework for building smart, secure and efficient systems." Journal of King Saud University - Computer and Information Sciences, 2018.
[http://dx.doi.org/10.1016/j.jksuci.2018.02.018]

[24]   X. Li, D. Zhang, L. Hu, and L.T. Yang, "AIoT: When Artificial Intelligence Meets the Internet of Things", *Mob. Netw. Appl.,* 2018.
[http://dx.doi.org/10.1007/s11036-018-1099-5]

[25]   A.J. Jara, L. Ladid, and A.F. Skarmeta, "Enabling the Internet of Things", *IEEE Commun. Mag.,* 2014.
[http://dx.doi.org/10.1109/MCOM.2014.6736742]

SUBJECT INDEX

www.ingramcontent.com/pod-product-compliance
Lightning Source LLC
Chambersburg PA
CBHW041444210326
41599CB00004B/123